Introduction to PSpice® Using OrCAD®

ELECTRIC CIRCUITS

NINTH EDITION

Introduction to PSpice® Using OrCAD®

ELECTRIC CIRCUITS

NINTH EDITION

JAMES W. NILSSON · **SUSAN A. RIEDEL**

Professor Emeritus, Iowa State University *Marquette University*

Prentice Hall

Boston · Columbus · Indianapolis · New York · San Francisco · Upper Saddle River
Amsterdam · Cape Town · Dubai · London · Madrid · Milan · Munich · Paris · Montreal · Toronto
Delhi · Mexico City · Sao Paulo · Sydney · Hong Kong · Seoul · Singapore · Taipei · Tokyo

Editorial Director, Computer Science and Engineering: *Marcia J. Horton*
Senior Editor: *Andrew Gilfillan*
Associate Editor: *Alice Dworkin*
Editorial Assistant: *William Opaluch*
Director of Marketing: *Margaret Waples*
Marketing Manager: *Tim Galligan*
Vice President, Production: *Vince O'Brien*
Senior Managing Editor: *Scott Disanno*
Production Editor: *Irwin Zucker*
Art Director: *Jayne Conte*
Art Editor: *Gregory Dulles*
Senior Operations Specialist: *Alan Fischer*
Marketing Assistant: *Mack Patterson*

Prentice Hall
is an imprint of

PEARSON

www.pearsonhighered.com

10 9 8 7 6 5 4 3
ISBN-13: 978-0-13-212307-5
ISBN-10: 0-13-212307-X

CONTENTS

Preface

ABOUT THIS MANUAL

Introduction to PSpice® Using OrCAD® 16.2 expressly supports the use of OrCAD PSpice A/D, Release 16.2 (herein after referred to as PSpice) as part of an introductory course in electric circuit analysis based on the textbook *Electric Circuits.* This supplement focuses on three things: (1) learning to draw and simulate linear circuits using PSpice, (2) constructing circuit models of basic devices such as op amps and (3) learning to challenge computer output data as a means of reinforcing confidence in simulation. Because PSpice is designed to simulate networks containing integrated circuit devices, its range of application goes well beyond the topics covered in the textbook. Even though we do not exploit the full power of PSpice, we begin the introduction to this widely used simulation program at a level that you can use to test the computer solutions.

The use of PSpice involves learning a new technical vocabulary and a number of specialized techniques. Hence, we designed this supplement to stand on its own as an instructional unit. Our decision to separate this material from the parent textbook also is a service to you: The portable format greatly facilitates your use of the supplement at a computer terminal. In this format, the supplement adds value to your study.

You may use PSpice to solve many of the textbook's Assessment Problems and Chapter Problems. Those Chapter Problems that we think are particularly suited to PSpice simulation are marked with a PSPICE icon in the textbook. You are encouraged to use PSpice to check your solutions to Chapter Problems, or to further explore the behavior of an interesting circuit.

Table 1: Relating Topics in this Supplement to Topics in the Text

MANUAL CHAPTER	TOPIC	TEXT CHAPTER
2–3	DC analysis	2–4
4	Thévenin Equivalents	4
4	Tolerance and Sensitivity	4
5	Op amps	5
6	Natural Response	7–8
7	Step response	7–8
7	Switches	7
8	Varying component values	8
9	AC Steady-state analysis	9
10	Linear Transformers	9–10
10	Ideal transformers	9–10
11	AC Power	10–11
12	Frequency response	13–15
12	Filter Design	14–15
13	Pulsed sources	16
13	Fourier series analysis	16

INTEGRATING PSPICE INTO INTRODUCTORY CIRCUITS COURSES

Although some circuits courses cover PSpice as an independent topic, many instructors prefer to integrate computer solutions with the course. To support such integration topics appear in this supplement in the same order in which they are presented in the text. Table 1 summarizes the relationship between the supplement and the textbook.

ABOUT PSPICE

SPICE is a computer-aided simulation program that enables you to design a circuit and then simulate the design on a computer. SPICE is the acronym for a Simulation Program with Integrated Circuit Emphasis. The Electronics Research Laboratory of the University of California developed SPICE and made it available to the public in 1975.

Many different software packages are available that implement SPICE on personal computers or workstations. Among them, OrCAD PSpice A/D, from Cadence is the most popular. PSpice's popularity can be attributed to many factors, including its user-friendly interface, extensions to SPICE that support modeling of digital circuits and much more, and its no-cost basic version. This manual focuses on how to use Release 16.2, and all the examples were produced using this release. If you are using a different version of PSpice, or another package that implements SPICE, your interaction with the software may differ from what you see in this supplement.

We limited our applications of PSpice to the types of circuit problems discussed in the textbook. Although PSpice is a general-purpose program designed for a wide range of circuit simulation—including the simulation of nonlinear circuits, transmission lines, noise and distortion, digital circuits, and mixed digital and analog circuits—here we discuss the use of PSpice only for dc analysis, transient analysis, and steady-state sinusoidal (ac) analysis. You should refer to the OrCAD PSpice A/D User's Guide for information on all of the features of PSpice that are not discussed in this supplement.

To find out more about how to obtain a demo version of OrCAD, go to

http://www.cadence.com/products/orcad/pages/downloads.aspx

ASSESSMENT

In order to assess your understanding of the material in each chapter of this manual, we have included a set of suggested problems from the Ninth Edition of *Electric Circuits* at the conclusion of each chapter. You are encouraged to solve these problems by hand, and then use PSpice to verify your solutions or to create plots that enable you to visualize the circuit behavior.

JAMES W. NILSSON
SUSAN A. RIEDEL

Chapter 1

A FIRST LOOK AT PSPICE

The general procedure for using PSpice consists of three basic steps. In the first step, the user describes the circuit to be simulated or analyzed by drawing it in schematic form. In the second step, the user specifies the type of analysis desired, and directs PSpice to perform that analysis. In the third and final step, the user instructs the computer to print or plot the results of the analysis. In this first look, we will carefully develop a simple example to illustrate how to perform these three steps.

1.1 DRAWING THE CIRCUIT

To begin, run the OrCAD Capture CIS program, and select the option File/New/Project, as shown in Fig. 1. This option will invoke the New Project dialog box, as shown in Fig. 2. You should select a meaningful project name and the desired directory for your project files. You should also select the Analog Mixed-Mode Project Wizard, as shown in Fig. 2, which will guide you through the rest of the process.

You can create your PSpice project based on an existing project, or create a blank project. The dialog for making this choice is shown in Fig. 3. When you exit this dialog, you are back in Capture, but now have a circuit layout grid and the Capture toolbar, as shown in Fig. 4.

Now you are ready to draw your circuit. Begin by selecting the menu option Place/Part, or by clicking on the second vertical toolbar button, as shown in Fig. 4. This will invoke the Place Part dialog. If you know the name of the library containing the part you want, highlighting the library name will display only the parts contained in that library. If the library

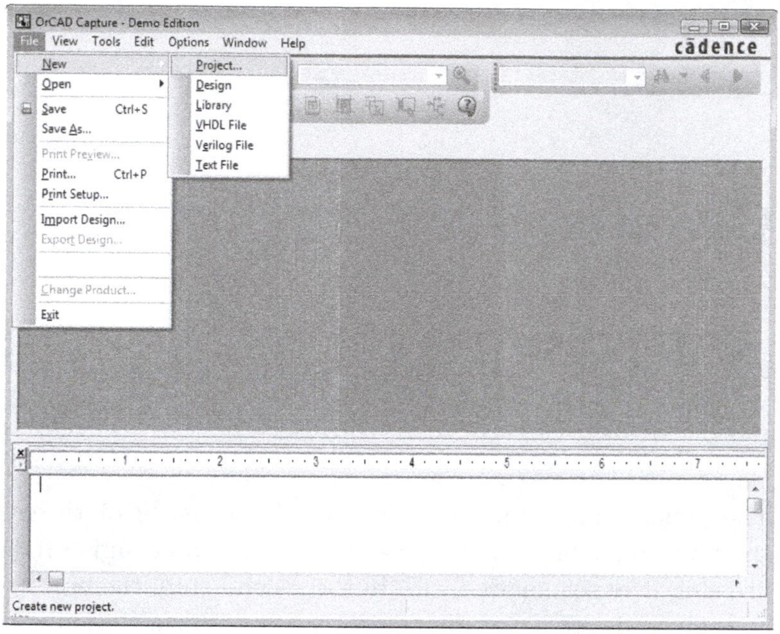

Figure 1: The initial window for OrCAD Capture CIS.

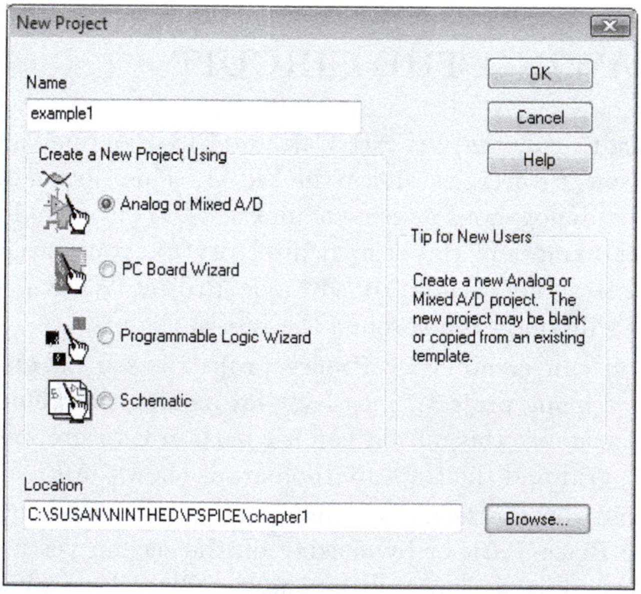

Figure 2: The New Project dialog box.

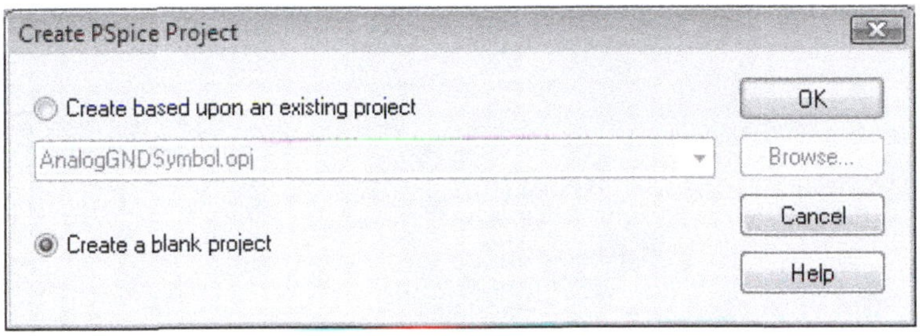

Figure 3: The dialog window for creating a PSpice project.

containing the part you need is not shown, you will need to add it by clicking on the Add Library button. The dialog for adding a library is shown in Fig. 5(a). Usually the libraries you need are in the Demo directory. As you can see in Fig. 5(b), we have selected the part named Vdc, a dc voltage source, from the library of sources named Source. If you are not sure of which library contains the part you need, you can highlight all of the libraries, and then all of the available parts will be displayed. When you click on a part name from the list its schematic is shown (see Fig. 5(b)), so you can verify that this is the part you want. Click on Place part (see Fig. 6) when you are ready to place the part in the schematic.

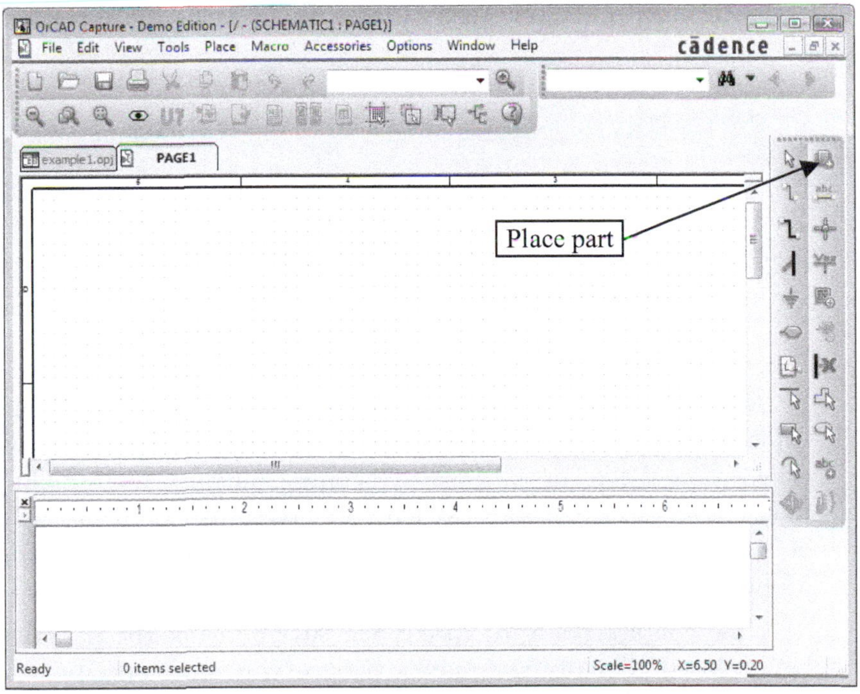

Figure 4: The circuit layout grid and Capture toolbar.

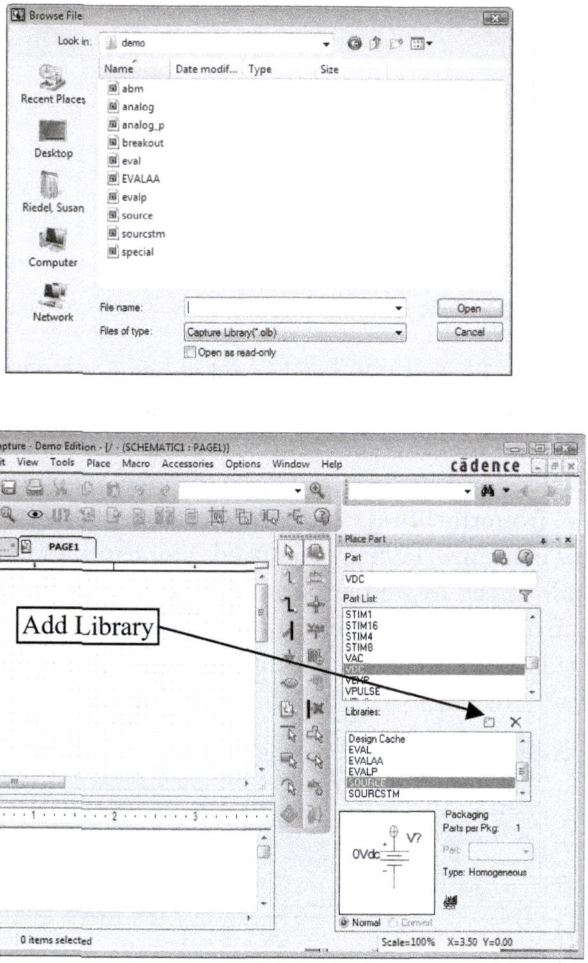

Figure 5: Selecting a part: (a) adding the library containing the part; (b) selecting the desired part.

Now you will be back in the Capture window and can place one or more copies of the part you have selected on the layout grid. You can change the orientation of the current part by typing Control-R or by right clicking the mouse and selecting rotate. When you have finished placing as many copies of the current part as you need, right click on the mouse and select End Mode, as shown in Fig. 6.

Now you are ready to specify the attributes of the part (or parts) you have placed. Usually the only attribute you must specify is the value of the part, which in this case is the value of the voltage supplied by the dc voltage source. To change this value, double click on it, and you will see the

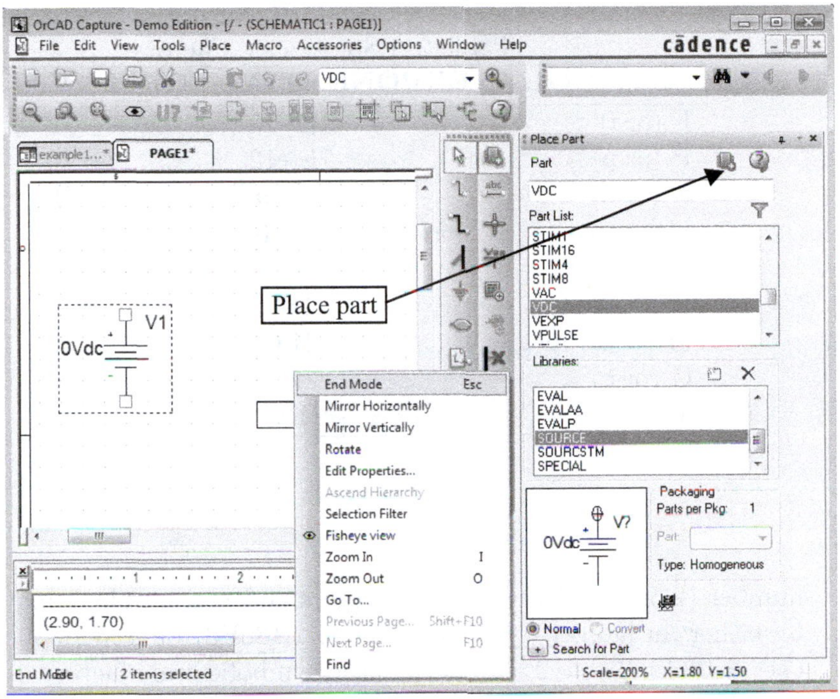

Figure 6: Place a dc voltage source on the layout grid, then right click on the mouse to finish with this part.

Display Properties dialog for this part, shown in Fig. 7. You can see from the figure that we have changed the value of the dc voltage source to 12V, or 12 volts. The number you specify may be an integer (4, 12, −8) or a real

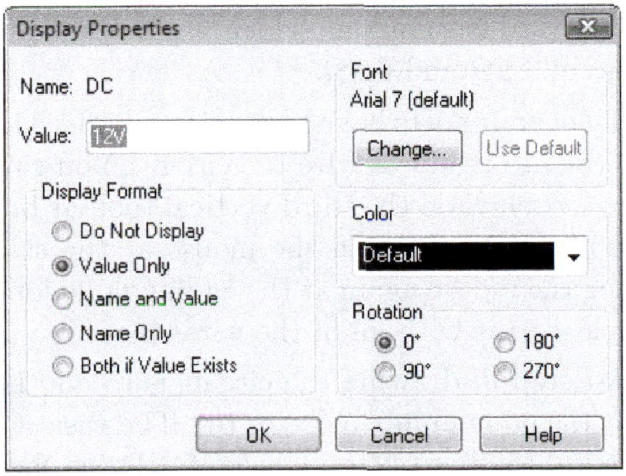

Figure 7: Using the Display Properties dialog window to set the value of the dc voltage source.

Table 2: PSpice Scale Factors

SYMBOL	EXPONENTIAL FORM	VALUE
F (or f)	1E−15	10^{-15}
P (or p)	1E−12	10^{-12}
N (or n)	1E−9	10^{-9}
U (or u)	1E−6	10^{-6}
M (or m)	1E−3	10^{-3}
K (or k)	1E3	10^{3}
MEG (or meg)	1E6	10^{6}
G (or g)	1E9	10^{9}
T (or t)	1E12	10^{12}

number (2.5, 3.14159, −1.414). Integers and real numbers may be followed by either an integer exponent (7E−6, 2.136E3) or a symbolic scale factor (7U, 2.136k). Table 2 summarizes the symbolic scale factors used in PSpice and their corresponding exponential forms. Letters immediately following a number that are not scale factors are ignored, as are letters immediately following a scale factor. For example, 10, 10v, 10HZ, and 10A all represent the same number, as do 2.5m, 2.5MA, 2.5msec, and 2.5MOhms. Note that you may not use any white space characters when you specify such values.

The process of locating a part, placing it on the layout grid, and setting the attribute values is repeated for each part in your circuit. Figure 8 shows the circuit we are working on, which has a 12 V dc source and two resistors with values of 3 kΩ and 6 kΩ.

Once all of your parts have been selected, placed, and had their attributes specified, you are ready to wire the circuit. You can select the menu option Place/Wire, or click on the third vertical toolbar button, as shown in Fig. 8. Wires are placed by clicking the mouse at the starting node for the wire, and clicking the mouse again at the ending node for the wire. Figure 9 shows the example circuit with all of the wires placed.

The last step in drawing the circuit is to add the circuit ground, which must be at the node numbered zero (0). The easiest way to do this is to click on the vertical toolbar button labeled GND (as shown in Fig. 9), and select the part named 0 in the CAPSYM library, as shown in Fig. 10. This is a very

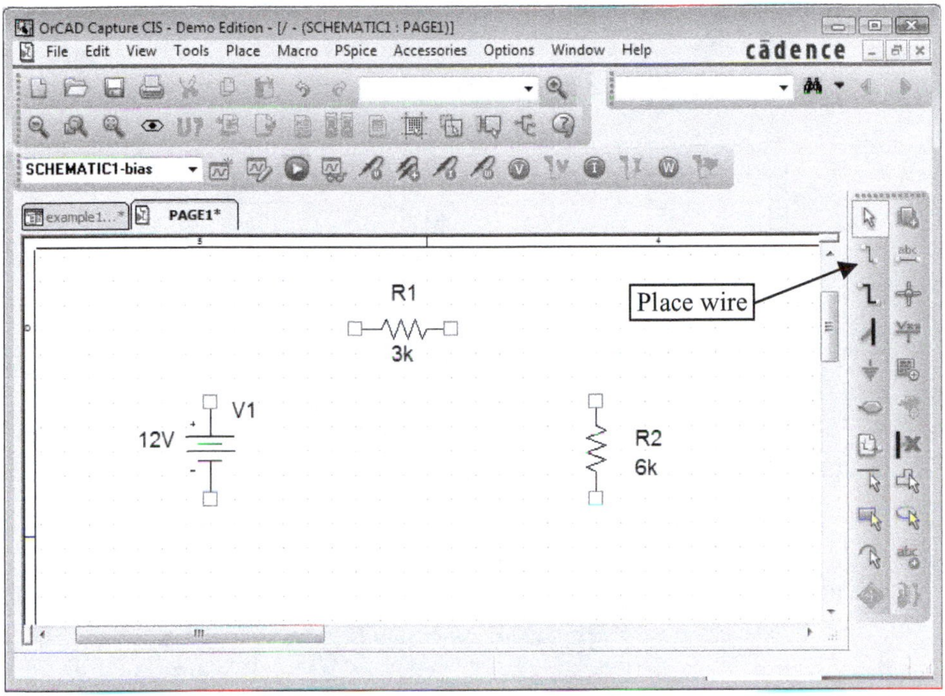

Figure 8: All of the circuit components have been selected and placed on the layout grid.

important step and easy to forget, but your circuit cannot be simulated by PSpice unless it contains a ground at node 0! Place the ground component on the layout grid, and wire it to your circuit. The completed circuit is shown in Fig. 11.

Now we are ready to specify the type of analysis to perform on this circuit. Before we specify the type of analysis to perform, let's briefly examine what PSpice will do with the circuit schematic we just drew. In order for PSpice to understand the circuit we have described, the schematic must be translated into a collection of statements that identify the circuit components, their attributes, and their topological connections. This collection of statements is called a netlist, and it is written to the output file before analysis begins. Sometimes it is useful to look at the netlist, especially if there is an error in the schematic. Therefore, the Appendix includes a subset of the netlist syntax, for your reference.

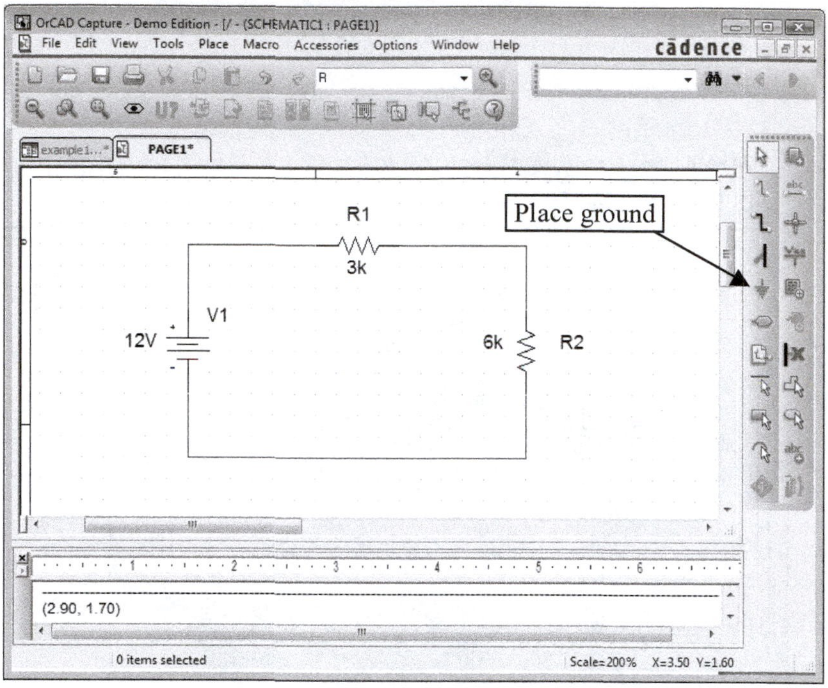

Figure 9: The circuit with all of the components wired together.

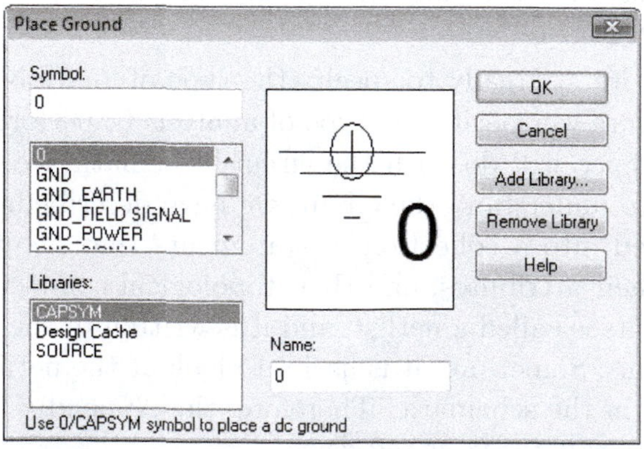

Figure 10: Selecting the ground component.

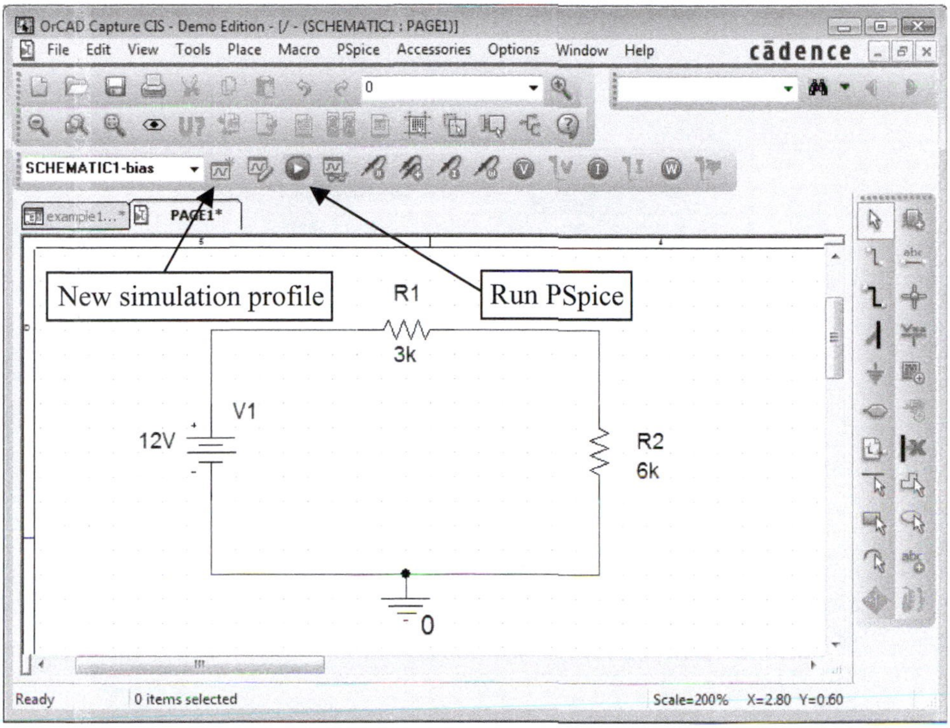

Figure 11: A completed circuit schematic.

1.2 SPECIFYING THE TYPE OF CIRCUIT ANALYSIS

You can ask PSpice to perform several different types of analysis on a circuit you have drawn. To begin, you need to select a name for the simulation. This is done by choosing the menu option PSpice/New Simulation Profile, or clicking the button indicated in Fig. 11, which will generate the dialog shown in Fig. 12. We've chosen the name "bias" for this profile, as we will

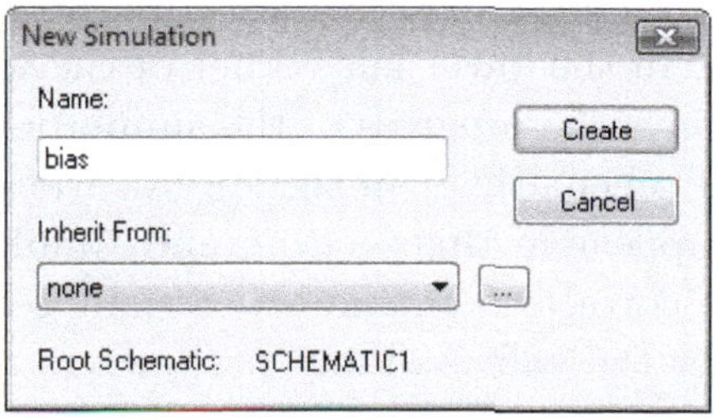

Figure 12: Creating a new simulation profile.

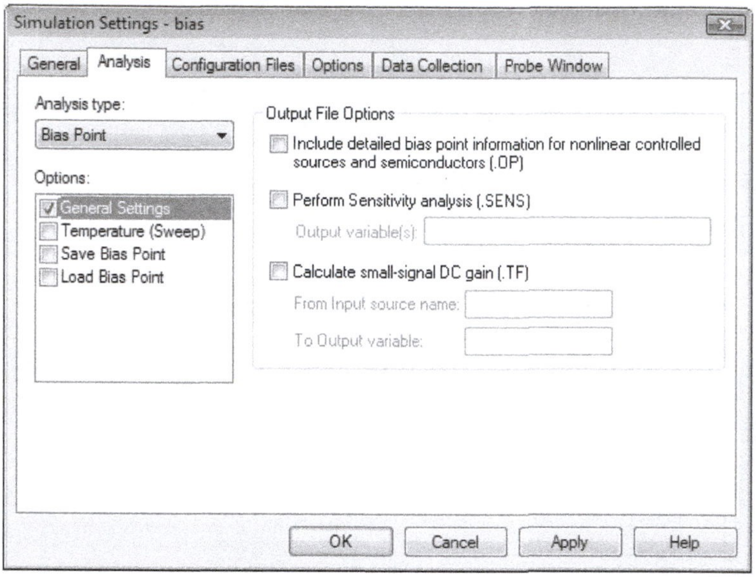

Figure 13: The Simulation Settings dialog window, used to specify the type of circuit analysis.

perform a simple dc analysis to determine all of the node voltage values, a process known as "biasing". Click the Create button and the Simulation Settings dialog window will appear, as shown in Fig. 13. Use the Analysis Type dropdown list on the Analysis tab to select the desired type of analysis. We have selected Bias Point, as you can see from Fig. 13. You can also see in this figure that there are many other tabs in the dialog which can be used to further configure the analysis. Usually, the default values suffice, as they do in this example. Once you select the analysis type, click on OK to return to your schematic.

Circuit analysis begins when you select the menu option PSpice/Run or by clicking the Run button shown in Fig. 11. You will see the PSpice A/D window, as shown in Fig. 14. As the figure shows, this window is divided into three sections. In the lower left corner of the window is an output text section that displays the progress of the simulation. If errors were encountered in your circuit schematic or in the specification of the analysis, a brief error message will appear in this section, and simulation will not continue until the errors are corrected. In the lower right corner is a section that allows you to monitor the values of any parameters that may be changing as the simulation progresses. There are no such parameters in this simple bias example. Finally, the top section can be used for printing or plotting results of the simulation, the topic of the next section.

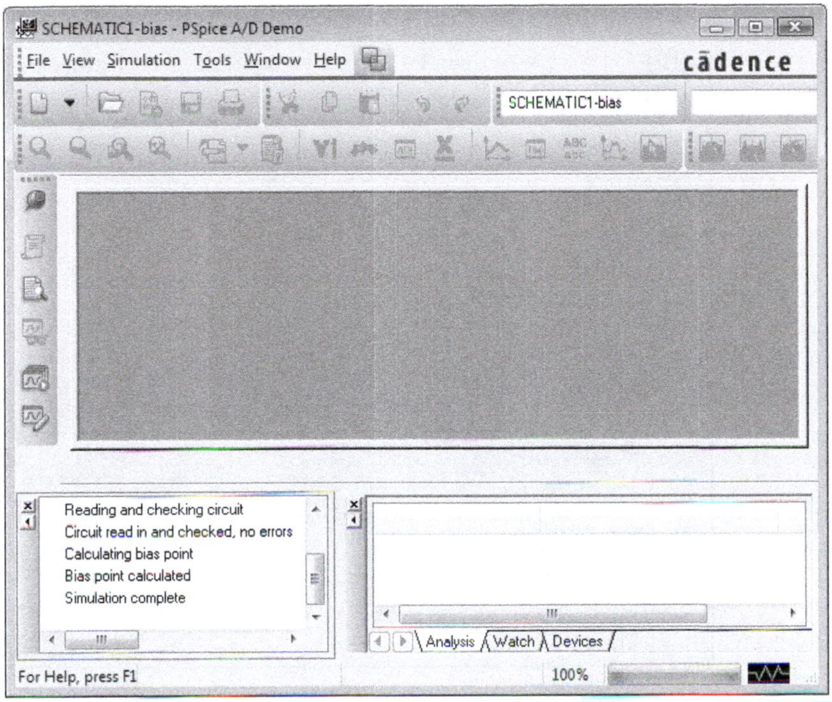

Figure 14: Circuit simulation using PSpice A/D.

1.3 SIMULATION RESULTS

Now we turn to the output from the simulation, which may be a text file, a plot, or both. In this simple example we cannot examine a plot, as no parameters have been varied. We examine the output from the bias analysis by clicking the third button on the left vertical toolbar. This opens the output file created during simulation in the top section of the PSpice A/D window, as shown in Fig. 15. Near the top of this output file is the netlist, the collections of statements that describe the circuit you represented as a schematic in a form that PSpice can understand. As you can see from the figure, the netlist consists of the three parts we drew: a 12 V dc voltage source between nodes N00727 and 0 (the required ground node); a 3 kΩ resistor between nodes N00727 and N00734; and a 6 kΩ resistor between nodes N00734 and 0.

If we scroll further down in the output file, we see the results of the bias analysis, including the value of the voltage at each of the non-ground nodes, as shown in Fig. 16(a). The voltage at node N00727 is 12 V, which we expected since the 12 V dc voltage source is between node N00727 and ground. The voltage at node N00734 is 8 V, which we also should have

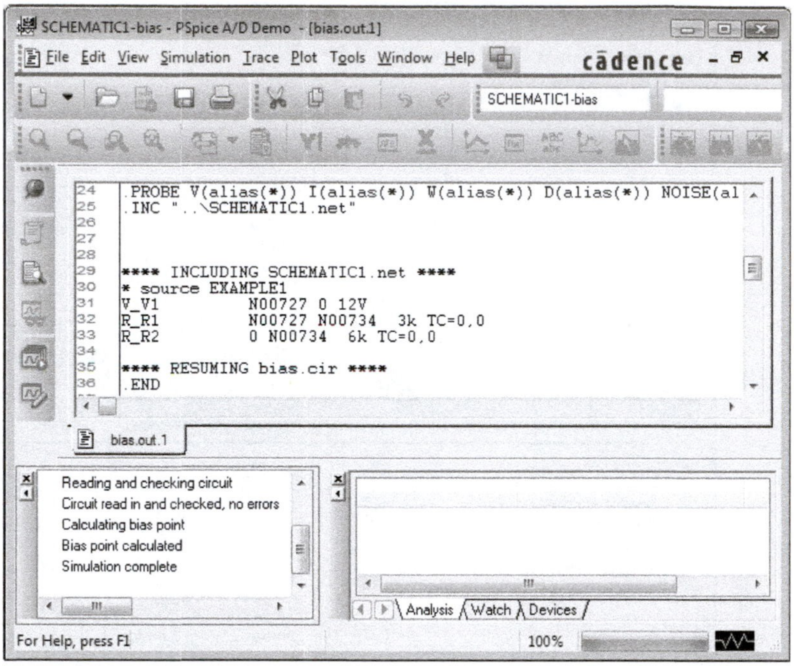

Figure 15: Displaying the output file produced during circuit simulation.

expected, since the circuit we created is a simple voltage divider and

$$v_{6\mathrm{k}\Omega} = \frac{6}{3+6}(12) = 8 \text{ V}.$$

We can also see these node voltage values displayed directly on the schematic. To do this, exit PSpice A/D and return to the schematic. Click on the V symbol in the second-row toolbar, and the node voltages will be displayed, as seen in Fig. 16(b).

This completes the simple example. The remainder of this supplement describes the various types of components used in linear circuits, the other types of analysis available in PSpice, and the use of Probe to plot the results of circuit simulation. Note that historically, the plotting functions in PSpice were collectively known as "Probe". We will continue to call these functions Probe in this text.

ASSESSMENT

- Use PSpice to solve the following problems: 3.11(a), 3.25(a), 3.27(a), and 3.28.

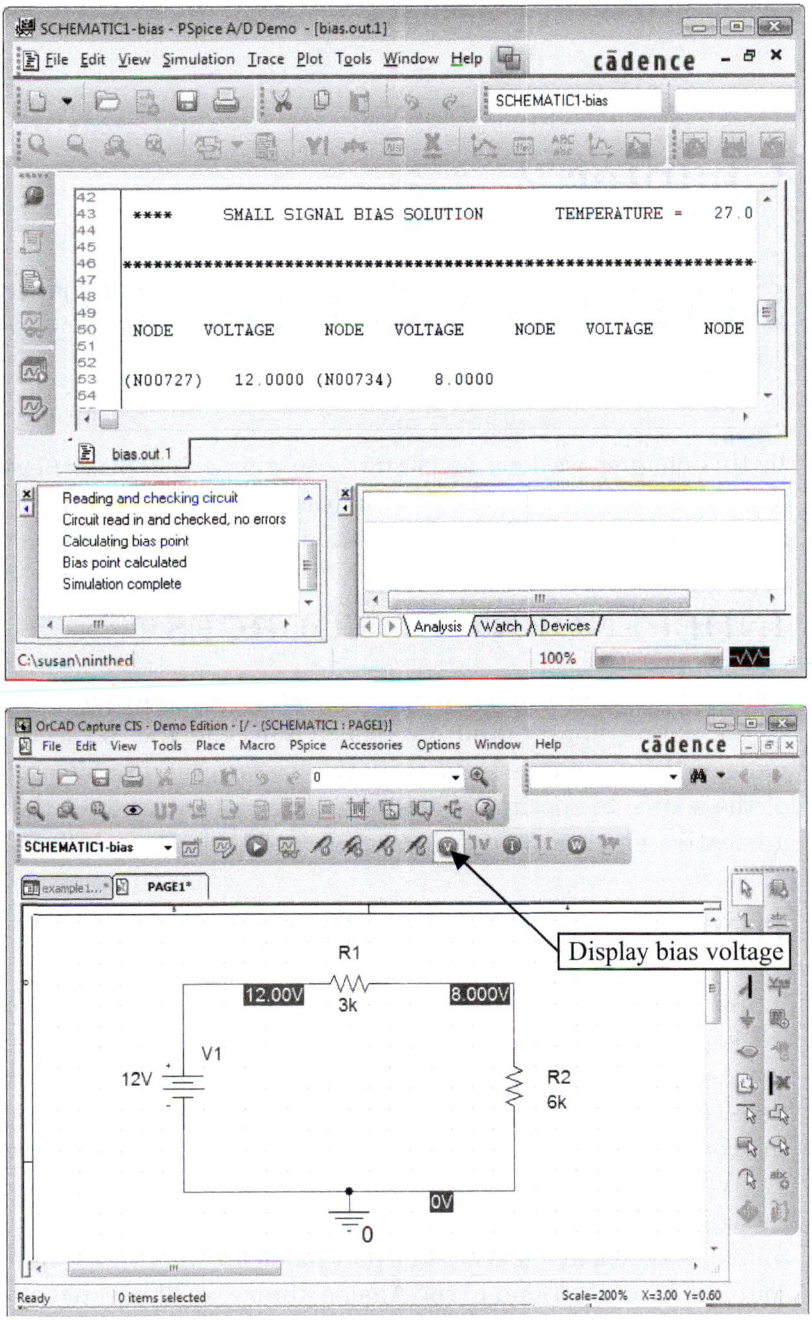

Figure 16: Results from the bias analysis: (a) the output file; (b) the schematic with node voltages displayed.

Chapter 2

SIMPLE DC CIRCUITS

In this chapter we look at circuits containing independent sources and/or dependent sources, together with resistors. We continue to analyze such circuits using a simple bias calculation.

2.1 INDEPENDENT DC SOURCES

Parts which model independent dc voltage and current sources are found in the Source library, and are named Vdc and Idc, respectively. Once the independent sources are placed in the layout grid, double click on the value of the source to specify the desired value, which can be either a positive or a negative number. We saw an example of Vdc in Chapter 1.

2.2 DEPENDENT DC SOURCES

In discussing the dependent dc sources, we divide them into voltage-controlled sources and current-controlled sources. Placing and wiring dependent sources can be a little tricky, as they each require four wired connections—two that connect the source into the circuit, and two that connect the controlling voltage or current into the source.

We start with a voltage-controlled voltage source. A circuit containing this type of dependent source, together with an independent dc current source, is shown in Fig. 17. The part name for the voltage-controlled voltage source is E, and is found in the Analog library, as illustrated in the Place Part dialog for this part, shown in Fig. 18. The schematic representation of this part, as seen in the figure, has four connection points. The two points on the right, labeled + and − inside a circle, are used to wire this component into

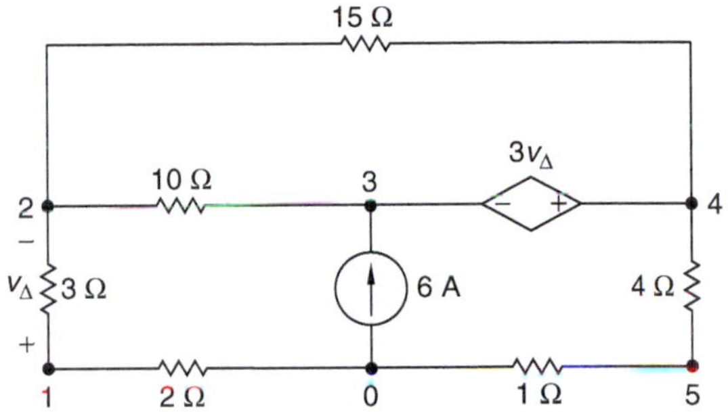

Figure 17: Circuit used to illustrate a voltage-controlled voltage source.

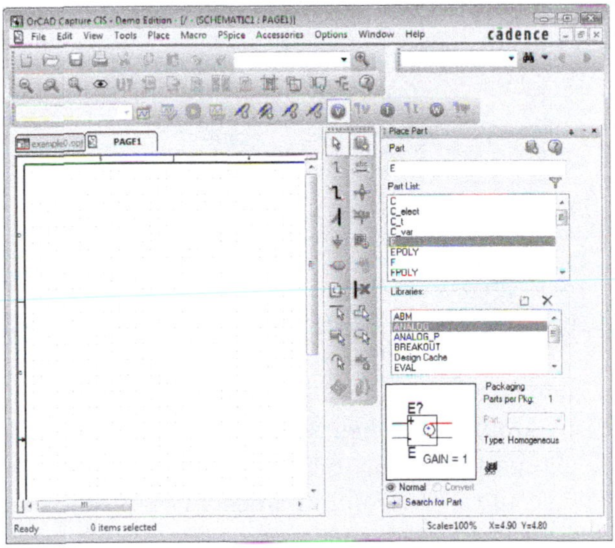

Figure 18: The Place Part dialog for the voltage-controlled voltage source.

the circuit between the 6 A current source and the 4 Ω resistor (see Fig. 17). The other two nodes are attached to a circuit to measure the controlling voltage. These two nodes must be wired into the circuit in parallel with the component across which the controlling voltage is defined. This is the 3 Ω resistor in Fig. 17, where the controlling voltage v_Δ is defined.

The circuit schematic with all of the parts placed but none of the connecting wires, is shown in Fig. 19. Note that we have rotated the voltage-controlled voltage source in preparation to orient it for wiring. Before we wire the schematic, we need to specify the gain of the voltage-controlled voltage source. To do this, click on the schematic of the voltage-controlled

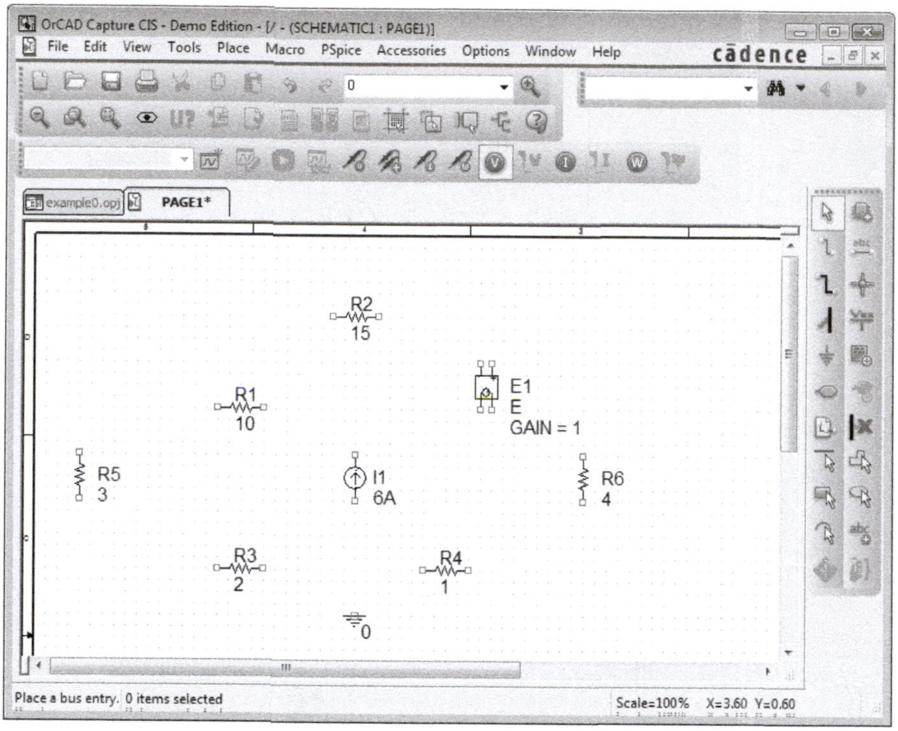

Figure 19: Parts placed for the schematic describing the circuit in Fig. 17.

voltage source (not its labels), which will highlight this part and place a dashed box around it. Then select the menu option Edit/Properties to invoke the Property Editor. Scroll horizontally in the property spreadsheet until you encounter the column labeled GAIN, click on the value of the gain and change it to 3, as shown in Fig. 20.

Now the parts can be wired and the ground added to complete the schematic, which is shown in Fig. 21. Be careful to wire the circuit so as to achieve the polarities shown in the circuit diagram. You might need to zoom in on the schematic as you wire in order to see the polarity markings on the dependent source. Use the zoom toolbar button or the menu option View/Zoom.

The schematic representations for the remaining dependent dc sources are shown in Fig. 22. These dependent dc sources are also in the Analog library. The current-controlled current source has the part name F, the voltage-controlled current source has the part name G, and the current-controlled voltage source has the part name H. Remember to wire in the dependent current sources using the terminals with the arrow inside of a circle, shown on the right of the schematic for parts F and G. The dependent

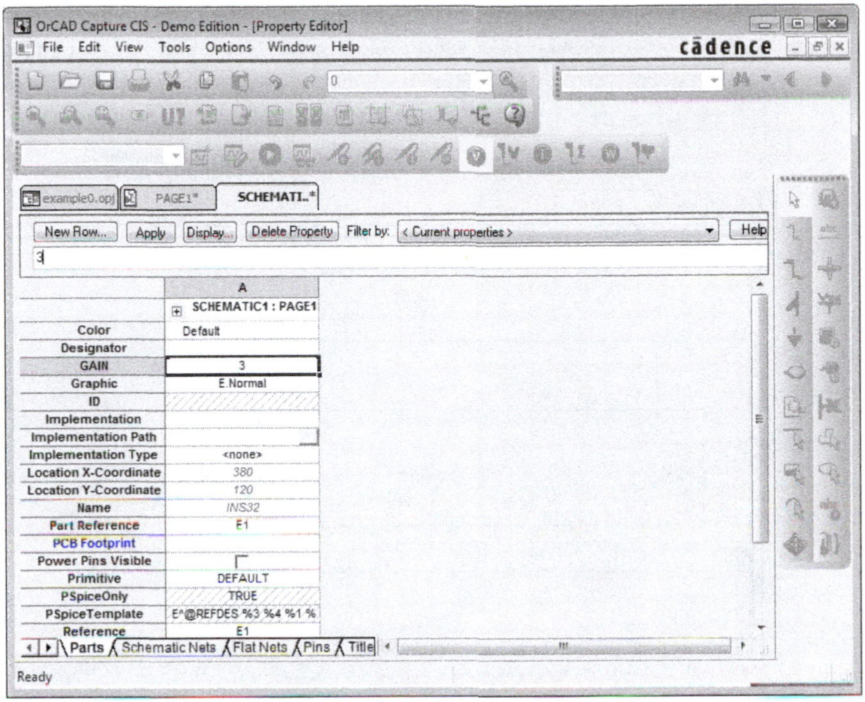

Figure 20: The Property Editor, used to change the gain of the voltage-controlled voltage source.

sources that are controlled by currents (parts F and H) have a large arrow on the left side of the schematic (see Fig. 22). These terminals must be wired in series with the component through which the controlling current is defined. We will see an example of a circuit employing a current-controlled dependent source in the next section.

2.3 RESISTORS

Chapter 1 performed a dc bias analysis on a circuit containing resistors and a dc voltage source. There, we saw that the model of a resistor has the part name R, and this part is found in the Analog library. Once we placed the resistors in our schematic, double clicking on their resistance value allows us to modify the resistance. The following example includes the bias point analysis of a circuit containing an independent source, a dependent source, and some resistors.

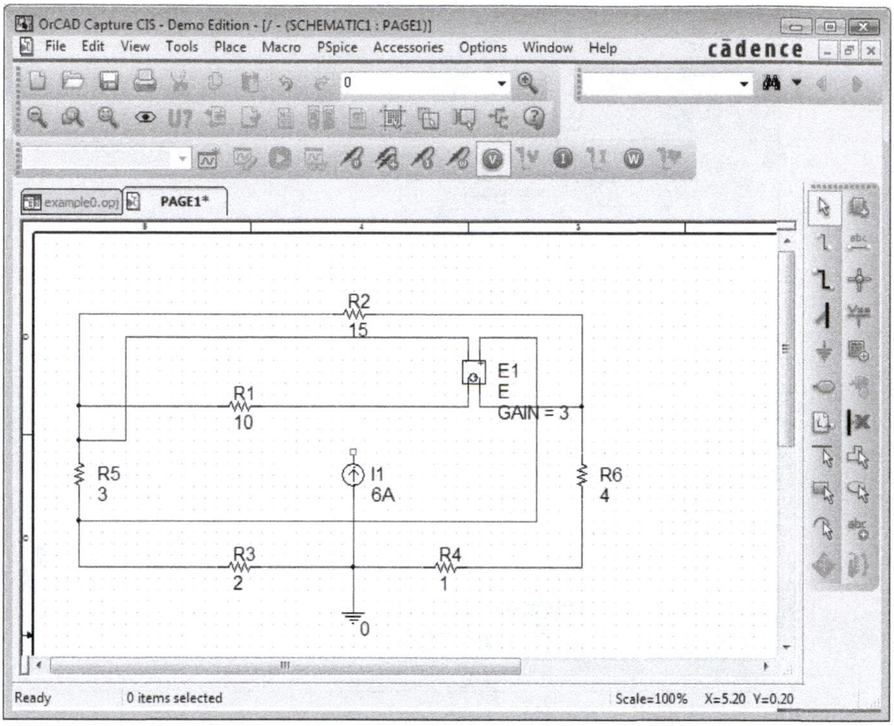

Figure 21: The completed schematic for the circuit in Fig. 17.

EXAMPLE 1

a) Use PSpice to find the voltages v_a and v_b for the circuit shown in Fig. 23. Enter the gain of the dependent source as 20.384615 (=265/13) in the Property Editor.
b) Use the PSpice solutions to calculate (1) the total power dissipated in the circuit, (2) the power supplied by the independent current source, and (3) the power supplied by the current-controlled voltage source.

SOLUTION

a) The circuit shown in Fig. 23 is represented as a schematic for PSpice as shown in Fig. 24. Note the wiring of the current-controlled voltage source. The pertinent PSpice output is shown in Fig. 25. From the PSpice output,

$$v_a = V(N00138) = 104.00 \text{ V};$$
$$v_b = V(N00248) = 106.00 \text{ V}.$$

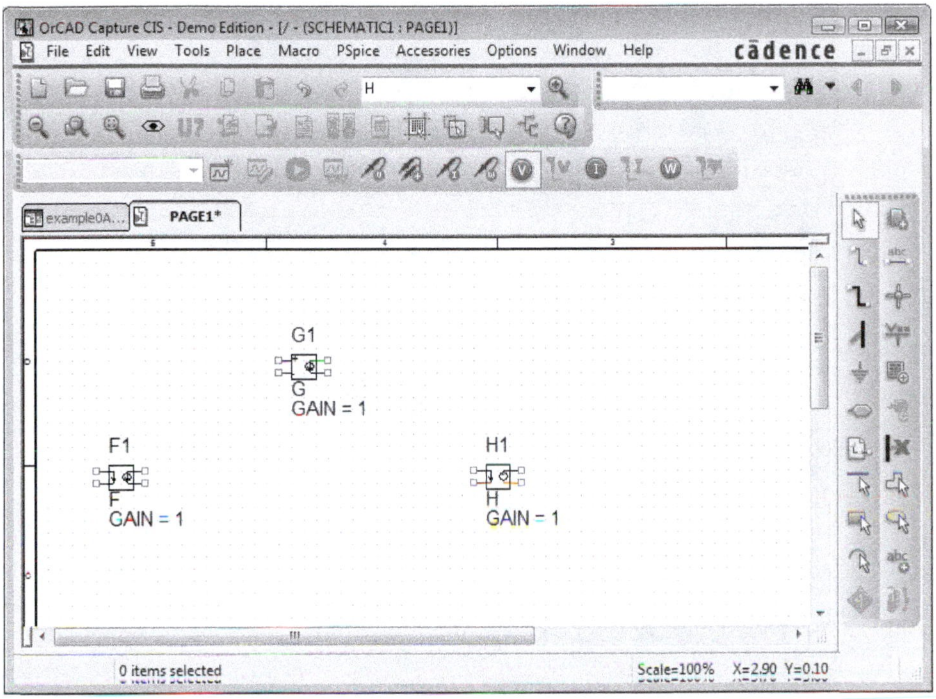

Figure 22: The schematic for the current-controlled current source (part name F), the voltage-controlled current source (part name G), and the current-controlled voltage source (part name H).

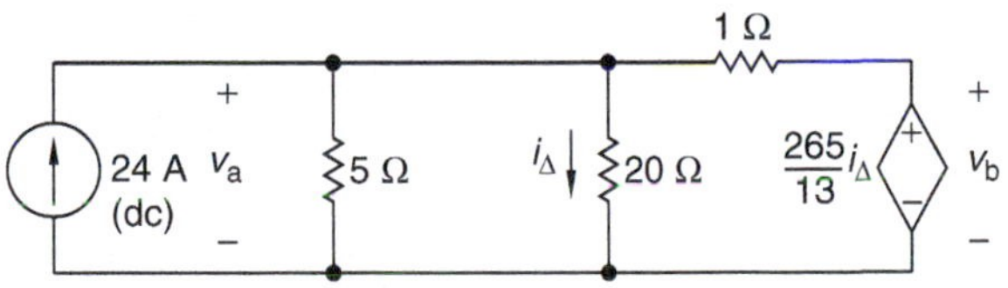

Figure 23: Circuit for Example 1.

b) The total power dissipation value is automatically printed whenever you invoke simple dc analysis. In PSpice, this power represents the net power generated by the independent voltage sources in the circuit. If a dependent source or an independent current source is generating power, its value is not included. That explains why the PSpice output in Fig. 25 gives the total power dissipation as 0 W—there are no independent voltage sources in the circuit that was simulated. Therefore, to find the actual value of the power dissipated and supplied, we need to write the following equations:

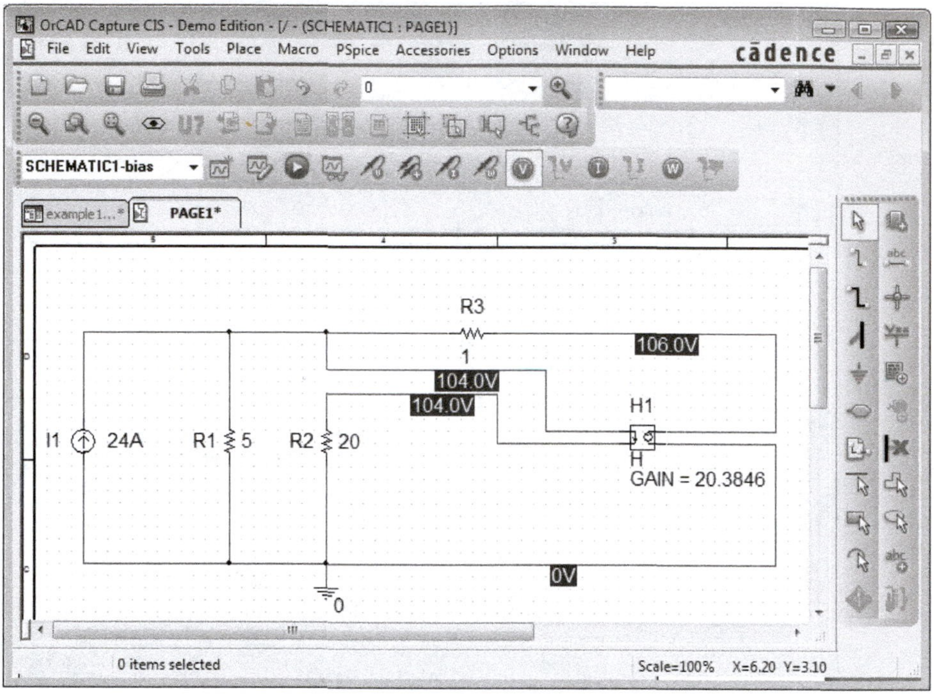

Figure 24: The schematic representation of the circuit shown in Fig. 23.

1.

$$\begin{aligned}
P_{5\Omega} &= \frac{104^2}{5} = 2163.20 \text{ W;} \\
P_{20\Omega} &= \frac{104^2}{20} = 540.80 \text{ W;} \\
P_{1\Omega} &= \frac{(106-104)^2}{1} = 4.00 \text{ W;} \\
\sum P_{\text{dis}} &= 2163.2 + 540.8 + 4 = 2708 \text{ W.}
\end{aligned}$$

2. $P_{24\text{A}}(\text{supplied}) = 104(24) = 2496$ W.

3. $P_{\text{H1}}(\text{supplied}) = 106(2) = 212$ W.

Note that the sum of power dissipated equals the sum of power supplied.

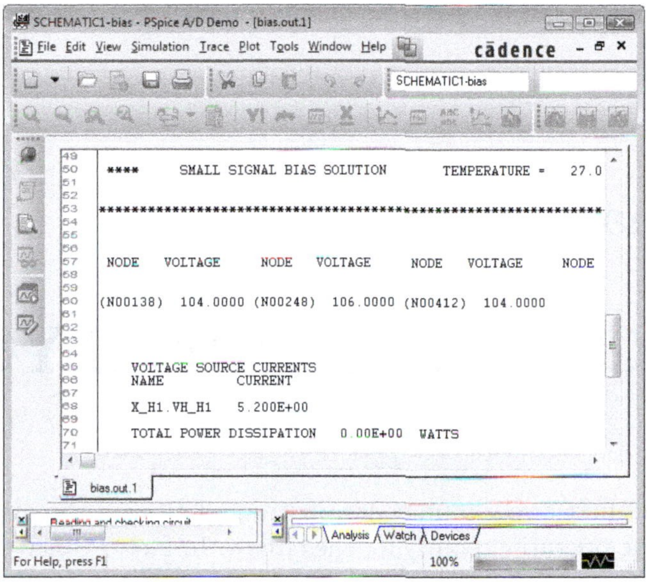

Figure 25: Part of the output file from the PSpice simulation of the circuit schematic shown in Fig. 24.

ASSESSMENT

- Use PSpice to solve the following problems: 4.6, 4.9, 4.15, 4.20, and 4.37.

Chapter 3

DC SWEEP ANALYSIS

In the first two chapters, we instructed PSpice to compute the bias point (the dc operating point) for the circuit. In fact, PSpice always calculates the operating point because this information is needed before you proceed with other types of analysis. As we saw, the result of bias point analysis was to print values in the output file. These values fall into three categories:

- voltages at each node
- currents in each voltage source and total power dissipated
- operating points for each element

In this chapter, we explore DC Sweep analysis, which enables us to simulate a circuit for a range of dc current or voltage source values.

3.1 SWEEPING A SINGLE SOURCE

Example 2 illustrates the use of DC Sweep analysis to examine the effect of different values of a voltage source on the output voltage and current in a circuit.

EXAMPLE 2

For the circuit shown in Fig. 26, use PSpice to find the values of i_o and v_o when v_g varies from 0 to 100 V in 10 V steps.

SOLUTION

The circuit shown in Fig. 26 is specified in schematic form in Capture, as shown in Fig. 27. Several new parts have been used in this schematic. The voltage source is modeled by the part Vsrc from the Source library, and the current source is modeled by the part Isrc from the Source library. Note that unneeded labels for these parts are not shown in the figure. We could also have used the parts Vdc and Idc for this example. To specify the value of the current source, highlight the part by clicking on it and select the menu option Edit/Properties. Scroll over to the column

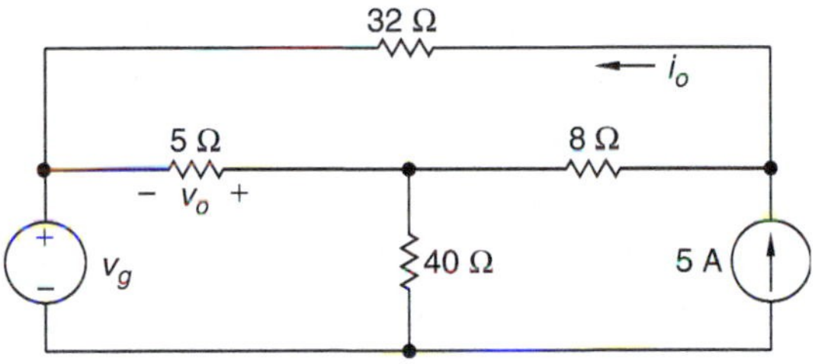

Figure 26: The circuit for Example 2.

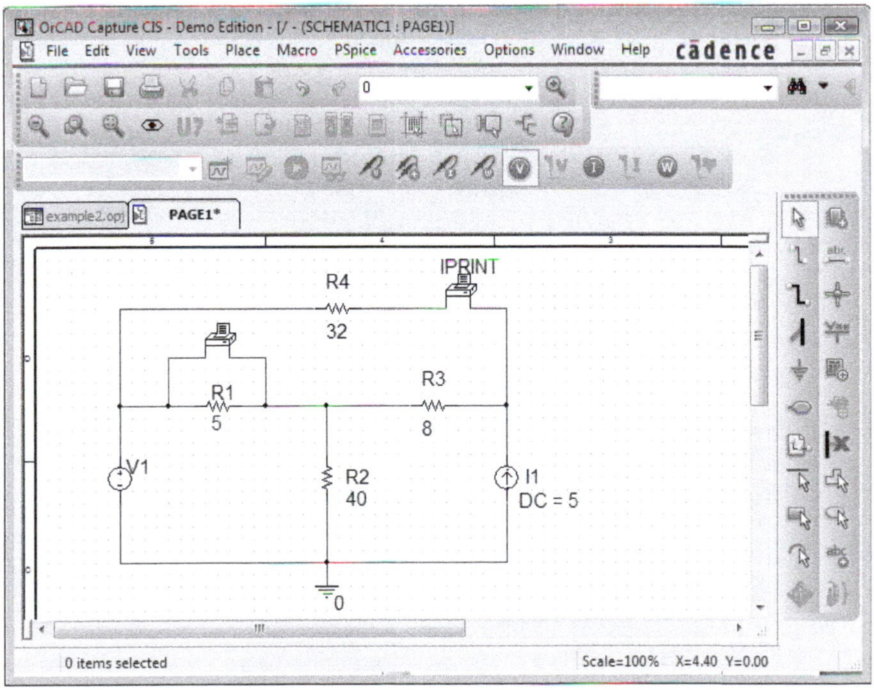

Figure 27: The circuit shown in Fig. 26 in schematic form.

labeled DC and type in the value of the current source in amps, 5. There is no need to specify the value of the voltage source, as you will see.

We have also added printers to the schematic, to cause the values of output current and output voltage to be written to the output file. The current printer has the part name Iprint, and is located in the Special library. The current printer must be placed in series with the branch through which the current of interest flows, so it acts like an ammeter. It is important to orient the current printer in the direction defined for the current. Therefore, place the symbol in the schematic so that the current exits the side of the symbol with the – sign (see Fig. 27). To tell the printer to print the results of dc analysis, highlight the printer and select the menu option Edit/Properties. Scroll over to the column labeled DC and type a Y in this entry, as shown in Fig. 28.

There are two voltage printer parts, both in the Special library. Vprint1 prints the voltage at a single node (with respect to ground), and Vprint2 prints the voltage between two nodes. We have used Vprint2

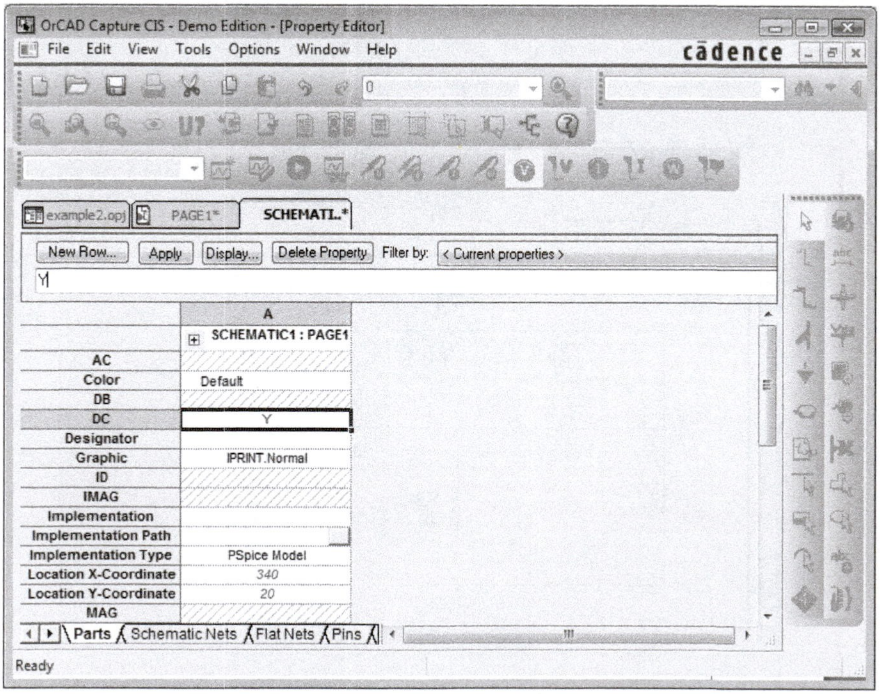

Figure 28: The Property Editor with the spreadsheet of properties for the current printer part Iprint.

in the schematic shown in Fig. 27. This part must be wired into the circuit in parallel with the component whose voltage is being measured, just like a voltmeter. Again, the orientation of this part is important, so place the symbol such that the node labeled – in the circuit schematic is wired to the side of the printer symbol with the – sign (see Fig. 27). You must edit the properties of the voltage printer and type a Y into the DC entry to print the voltage values in the output file.

Once the schematic is complete, create a new simulation profile for this circuit and select DC Sweep as the analysis type. Enter the name of the part whose values are to be swept, in this case the voltage source V1, and the Start Value, Final Value, and Increment, as shown in the Simulation Settings dialog in Fig. 29. Run the PSpice simulation, and look at the output file, a portion of which is shown in Fig. 30, where you can see the effect of changing the source voltage on the output voltage.

We can also depict the results of the DC Sweep analysis graphically. In fact, after PSpice has completed the simulation, you will notice that a graph is displayed in the upper section of the simulation window, with the value of the source voltage already selected as the independent variable. Use the Trace/Add Trace menu option to specify which variable

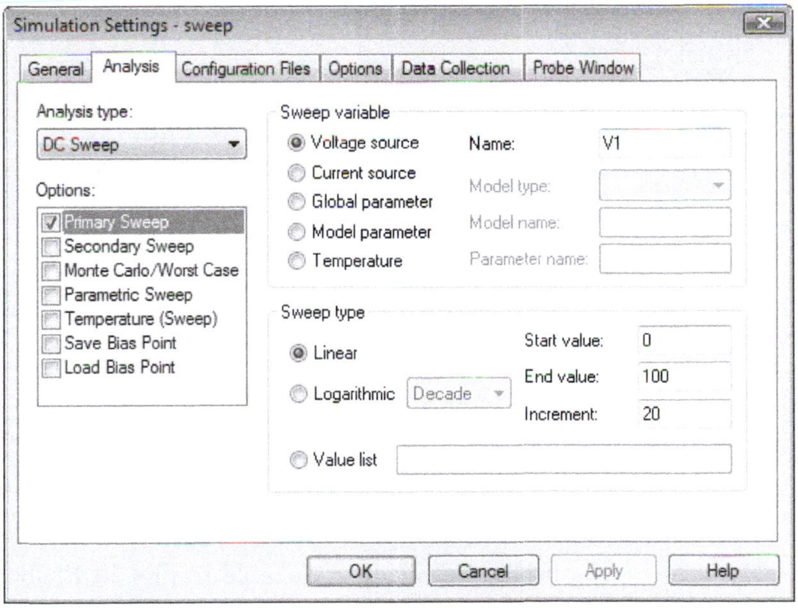

Figure 29: The Simulation Settings dialog for the DC Sweep analysis of the schematic in Fig. 27.

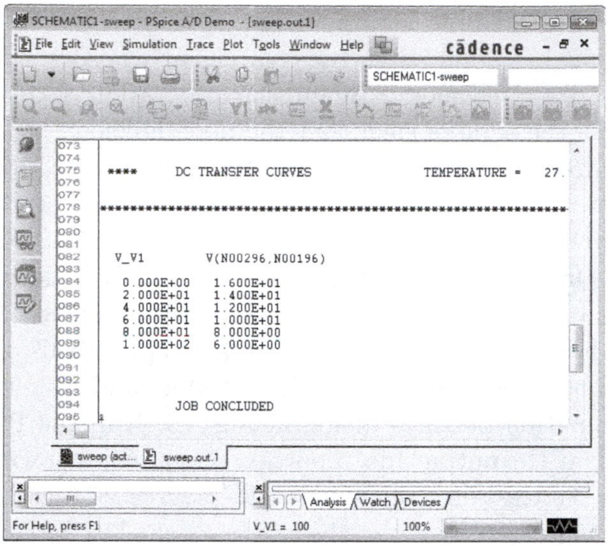

Figure 30: The results of DC Sweep analysis of the schematic in Fig. 27.

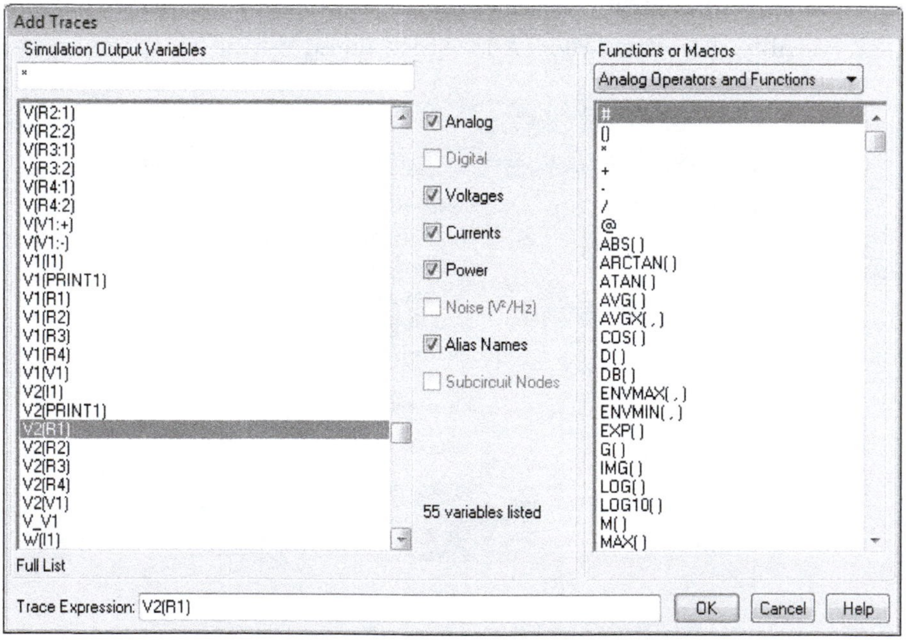

Figure 31: Selecting a variable to plot in Probe.

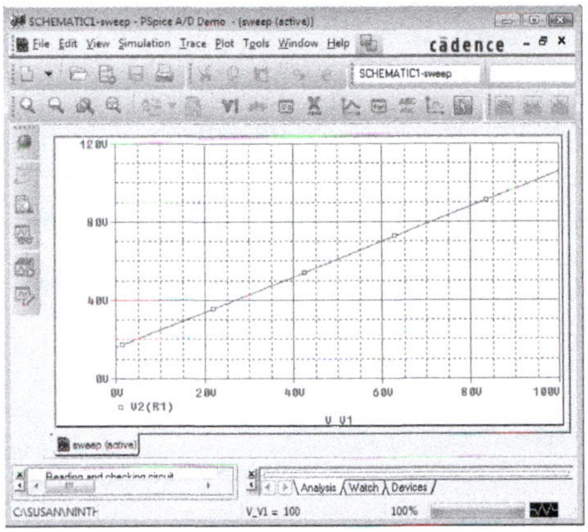

Figure 32: The plot of the output voltage versus the source voltage for the schematic in Fig. 27.

to plot. As shown in Fig. 31, we selected the voltage difference across the resistor R1, using the variable name V2(R1). The resulting plot is shown in Fig. 32.

3.2 SWEEPING MULTIPLE SOURCES

Now we alter the source file of Example 2 to sweep through values of the current and voltage sources. Example 3 presents the results.

EXAMPLE 3

The current source in the circuit shown in Fig. 26 varies from 0 to 5 A in 1 A steps. For each value of the source voltage, plot i_o as v_g varies from 0 to 100 V in steps of 20 V.

SOLUTION

The schematic is the same as the one in Fig. 27. In Capture, select the menu option PSpice/Edit Simulation Profile to specify a Secondary Sweep, the additional sweep variable and its values. The Simulation Settings dialog is shown in Fig. 33. The resulting plot is shown in Fig. 34. There are six separate traces on this plot, one for each value of source current.

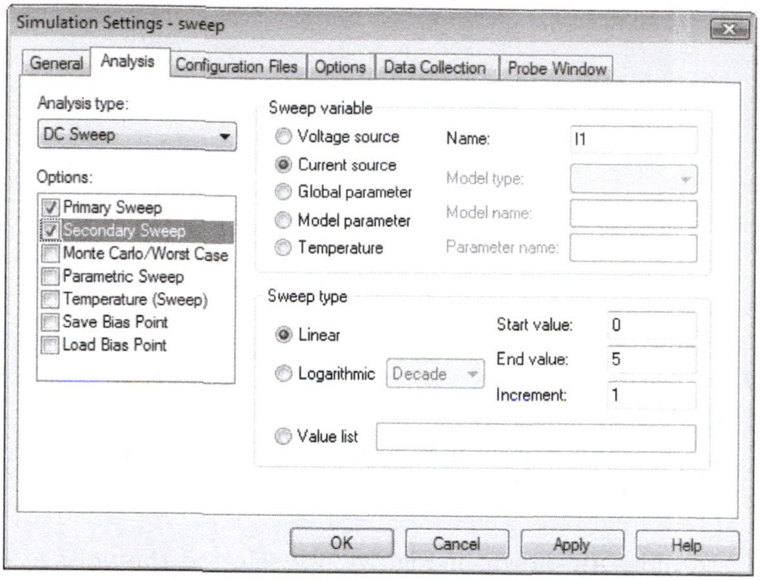

Figure 33: Adding a secondary sweep to the DC Sweep analysis of the schematic in Fig. 27.

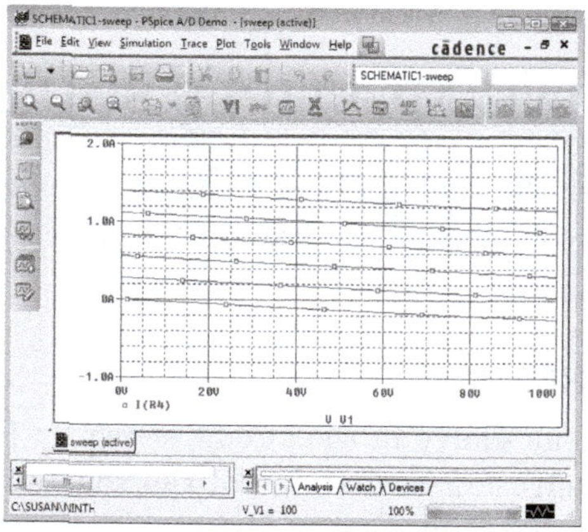

Figure 34: Plot of output current as source voltage and source current are varied, for the schematic in Fig. 27.

ASSESSMENT

- Use PSpice to solve the following problems: 4.57 and 4.58.

Chapter 4

ADDITIONAL DC ANALYSIS

This chapter expands the use of dc analysis to include using PSpice to compute a Thévenin equivalent, examining the sensitivity of certain circuit output variables to changes in circuit parameter values, and calculating the effect of resistor tolerances on circuit outputs.

4.1 COMPUTING THE THÉVENIN EQUIVALENT

You may use bias analysis to calculate the open-circuit voltage and the short-circuit current at a designated pair of nodes in a circuit, and thus compute the Thévenin equivalent with respect to that pair of terminals. Example 4 illustrates this application.

EXAMPLE 4

Use PSpice to find the Thévenin equivalent with respect to terminals a, b for the circuit shown in Fig. 35.

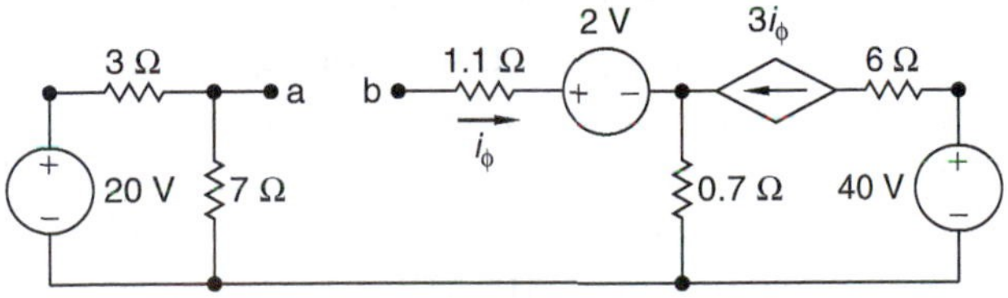

Figure 35: The circuit for Example 4.

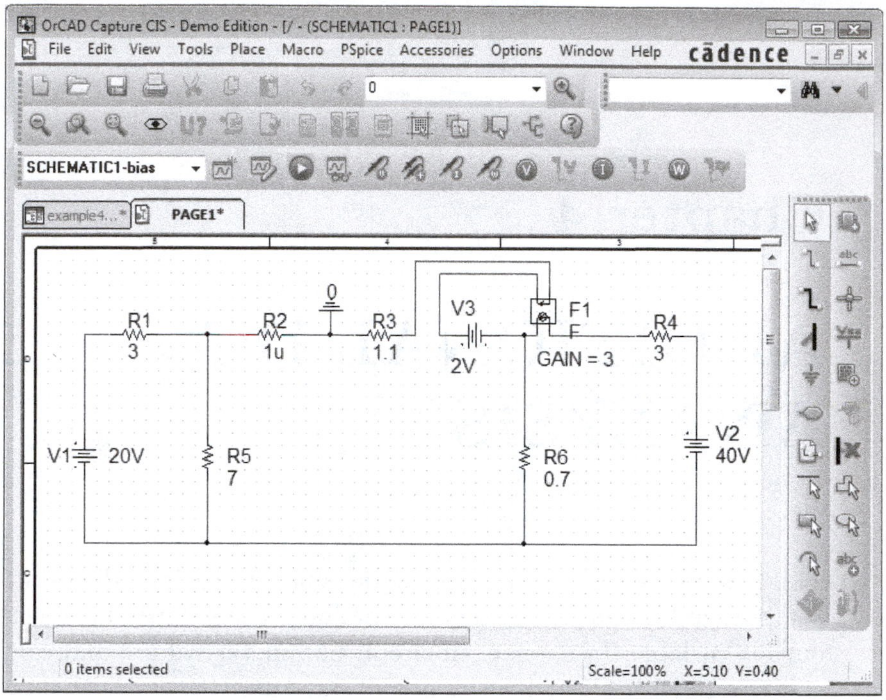

Figure 36: The schematic for the circuit in Fig. 35, configured to calculate the short-circuit current.

SOLUTION

A schematic created from the circuit shown in Fig. 35 is given in Fig. 36. Note that we have selected the node labeled b in Fig. 35 as the reference node. To compute the short-circuit current, we inserted a resistor between nodes a and b (node 0) whose value is much smaller than the value of other resistors in the circuit, in this case 10^{-6} Ω, effectively creating a short circuit. Bias analysis will automatically print the current through the 2 V source, which will be the short circuit current. The results of PSpice analysis, shown in Fig. 37, confirm that the short circuit current, which is the current through the voltage source V3, is 2 A.

To compute the open-circuit voltage, we cannot merely compute the voltage between nodes a and b in the circuit in Fig. 35 since PSpice requires at least two connections to every node. We have two options. First, we may connect a resistor that is much larger than the other resistors in the circuit—say, 10^6 Ω for this circuit—between nodes a and b, effectively creating an open circuit. Second, we may connect a capacitor between nodes a and b. The capacitor behaves like an open circuit during dc analysis and, therefore, does not influence the dc Thévenin equivalent.

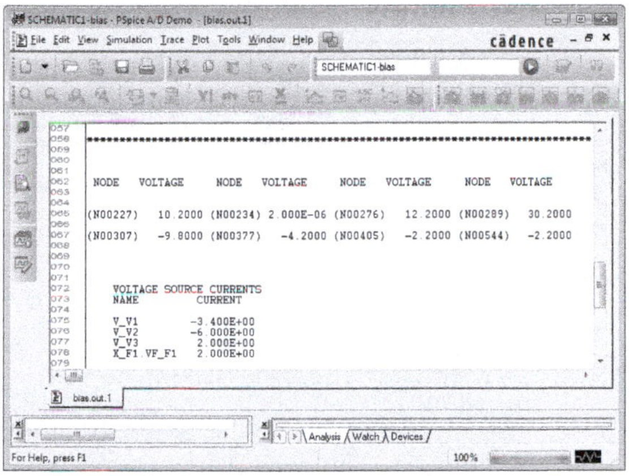

Figure 37: PSpice output calculating the short circuit current.

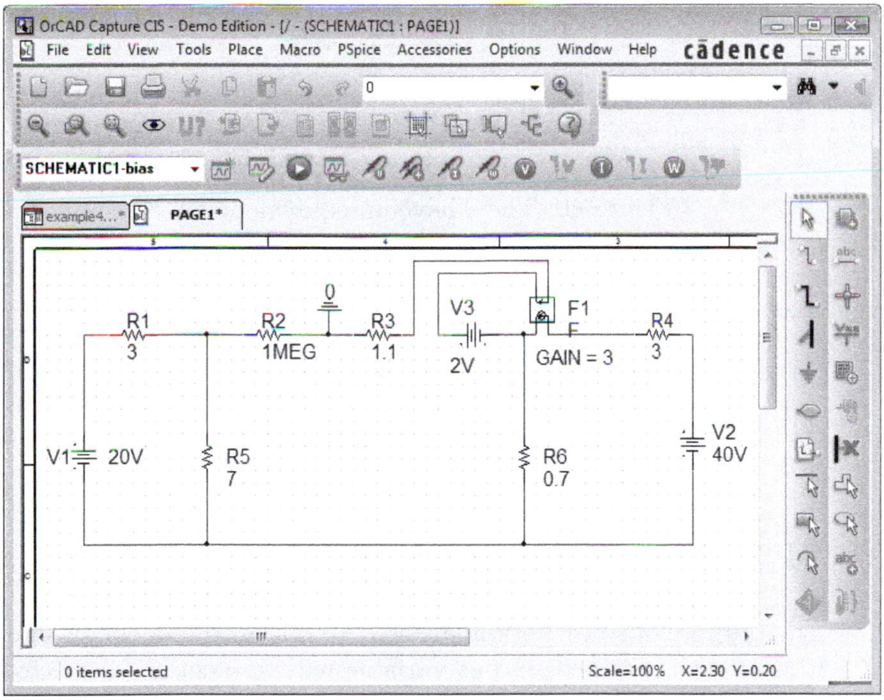

Figure 38: The schematic for the circuit in Fig. 35, configured to calculate the open-circuit voltage.

Here we use the first option, with the resulting schematic shown in Fig. 38. The pertinent data from the PSpice output from the simulation of the circuit in Fig. 38 is shown in Fig. 39. Note that node a in the

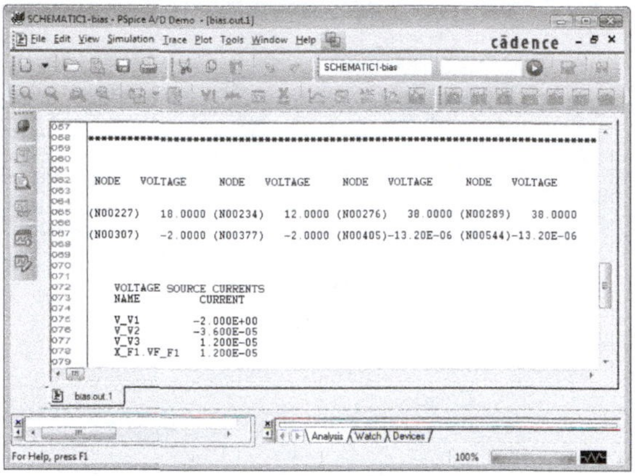

Figure 39: PSpice output file for determining the open-circuit voltage.

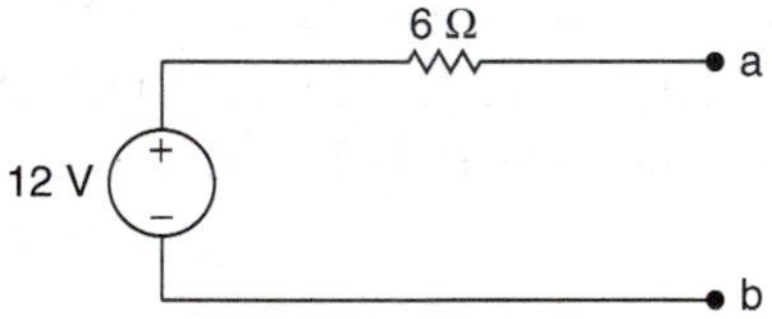

Figure 40: The Thévenin equivalent for Example 4.

original circuit (Fig. 35) is assigned as node N00234 by PSpice, so the open-circuit voltage is 12 V. Therefore, the Thévenin resistance is 12V/ 2A = 6 Ω and the Thévenin equivalent circuit is as shown in Fig. 40.

4.2 SENSITIVITY ANALYSIS

The purpose of sensitivity analysis is to obtain the dc small-signal sensitivities of each specified output variable with respect to *every* circuit parameter. Hence for circuits containing a large number of elements, tremendous amounts of output data can be generated, particularly when the sensitivity of more than one output variable is of interest. Sensitivity analysis can be performed as part of bias point analysis, and requires you to identify the voltages or currents in the circuit whose sensitivities are to be calculated. Example 5 illustrates how to perform sensitivity analysis to predict the behavior of an unloaded voltage-divider circuit.

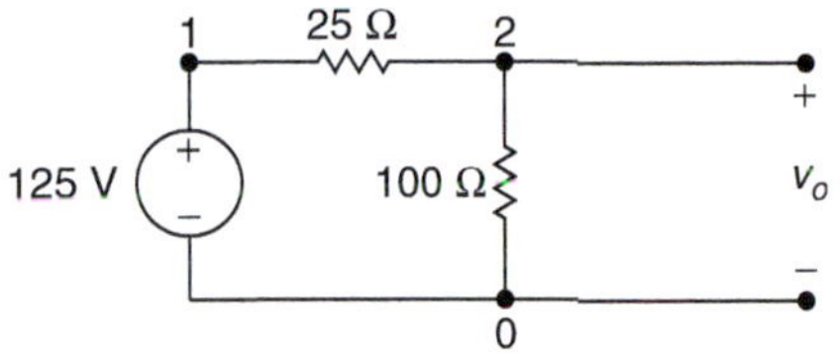

Figure 41: The circuit for Example 5.

EXAMPLE 5

Use PSpice to study the sensitivity of the output voltage v_o in the voltage-divider circuit shown in Fig. 41.

SOLUTION

Because we have three elements in the circuit, PSpice calculates the sensitivity of v_o with respect to the independent voltage source, the 25 Ω resistor, and the 100 Ω resistor. The PSpice schematic is shown in Fig. 42. Note that we have identified and labeled the output node of

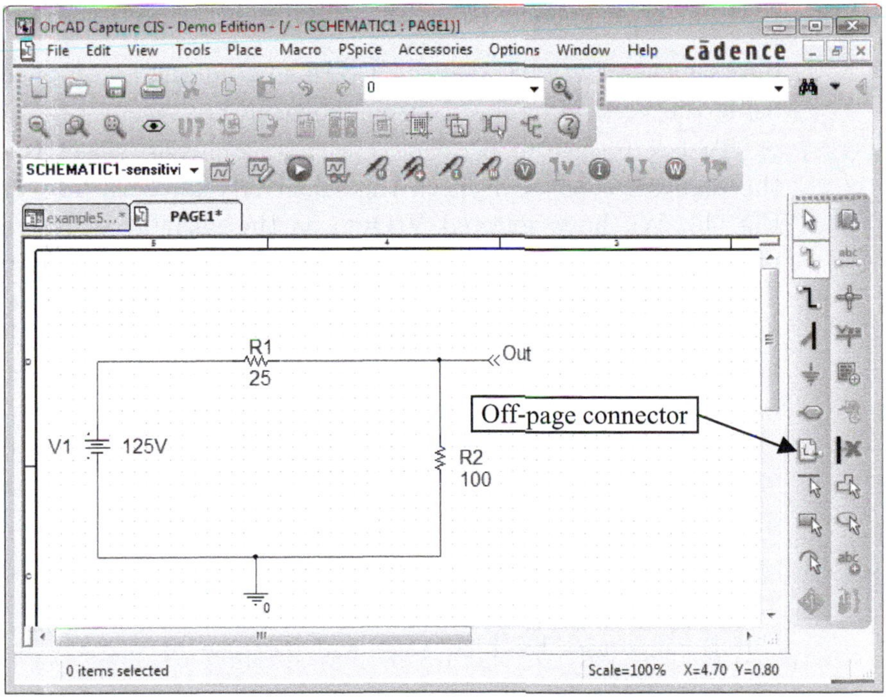

Figure 42: The PSpice schematic for the circuit in Fig. 41.

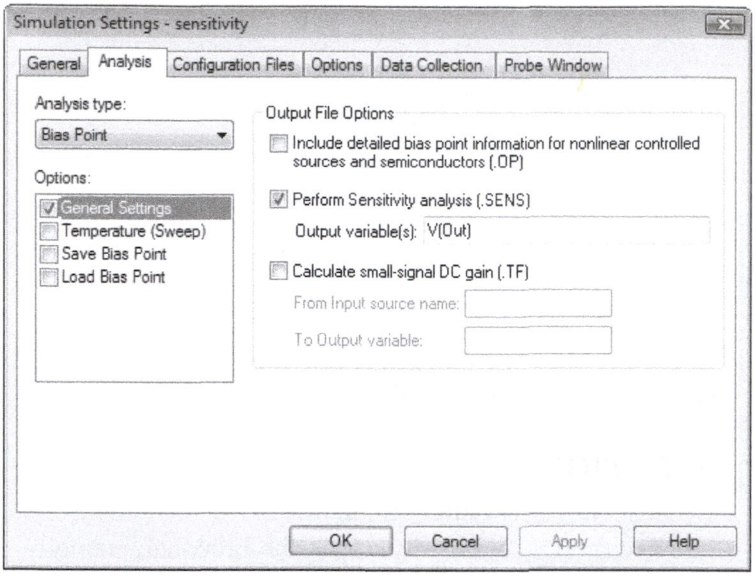

Figure 43: Specifying sensitivity analysis in PSpice.

interest for sensitivity analysis. Do this by adding an off-page connector using the vertical toolbar button shown in Fig. 42.

To ask PSpice to perform sensitivity analysis, select Bias Point as the analysis type, and check the sensitivity analysis box, as shown in Fig. 43. We have selected V(Out) as the variable of interest. PSpice prints the data associated with sensitivity analysis in the output file. The data relevant to sensitivity are shown in Fig. 44. From the simple dc analysis of the circuit, v_o = V(Out) = 100 V when R1, R2, and V1 are at their nominal values. From the sensitivity data, we deduce that

1. If R1 increases by 1 Ω, V(2) will decrease by 0.8 V—that is, V(2) = v_o = 99.2 V;

2. If R1 increases by 1%, v_o will decrease by 0.2 V to 99.8 V;

3. If R2 increases by 1 Ω, V(2) will increase by 0.2 V—that is, V(2) = v_o = 100.2 V;

4. If R2 increases by 1%, v_o will increase by 0.2 V to 100.2 V;

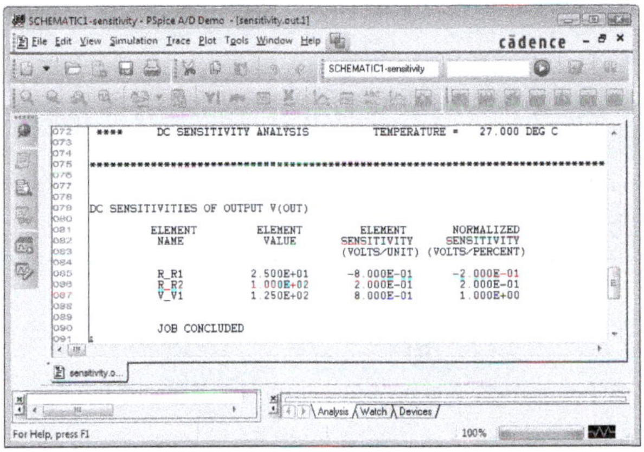

Figure 44: Part of the output file generated by simulating the circuit shown in Fig. 42.

5. If V1 increases by 1 V, V(2) will increase by 0.8 V—that is, V(2) $= v_o = 100.8$ V; and

6. If V1 increases by 1%, v_o will increase by 1 V to 101 V.

Because we have a linear circuit, the principle of superposition applies, and therefore we may superimpose simultaneous effects. For example, let's assume that R1 increases by 1 Ω, R2 decreases by 1%, and V1 increases by 0.5 V. The effect on v_o, or V(2), will be $v_o = 100 - 0.8 - 0.2 + 0.4 = 99.4$ V.

4.3 SIMULATING RESISTOR TOLERANCES

Sensitivity analysis enabled us to examine the effect of choosing a particular component value on the output of the circuit. But we might also be interested in simulating the effect of variation in component values, called tolerances, on the output of the circuit. In the case of resistors, the PSpice model used assumes that the resistor is perfect, never varying from its desired resistance value. In the next example, we alter the model of the resistor to specify a realistic tolerance value. With this more realistic component in our schematic, we then perform worst-case analysis to discover the impact of this tolerance on the output of our circuit.

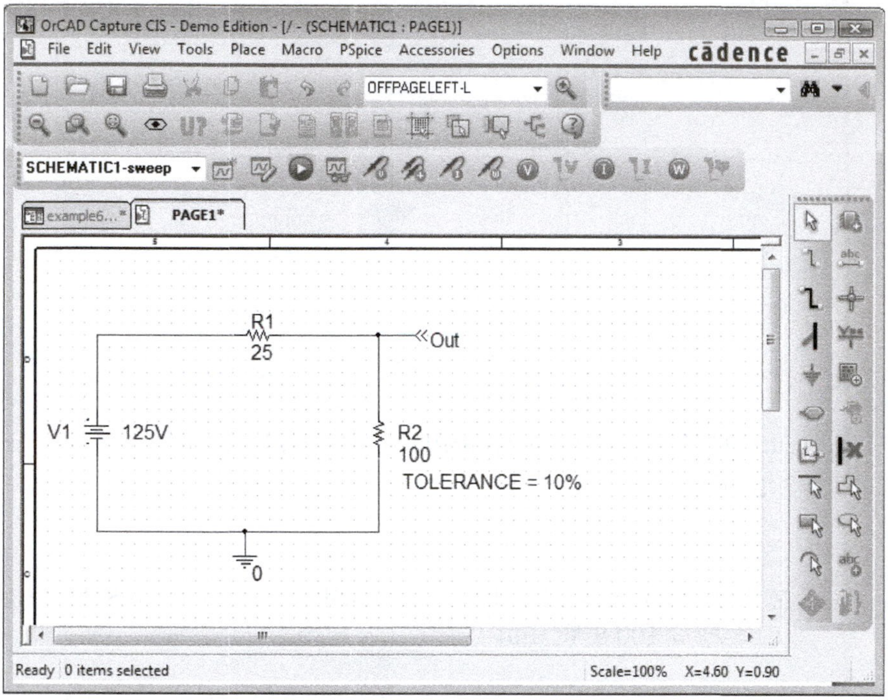

Figure 45: The schematic for the circuit in Fig. 41, using a resistor model with 10% tolerance.

EXAMPLE 6

Replace the 100 Ω resistor in the circuit in Fig. 41 with a resistor having the same value, but with a 10% tolerance. Use PSpice to discover the range of output voltage values that you can expect with this more realistic model of a resistor.

SOLUTION

The schematic for the circuit in Fig. 41 is shown in Fig. 45. Note that the R2 resistor displays 10% as the tolerance value. We set this tolerance value using the Property Editor to access the spreadsheet of properties for resistors. Find the column labeled Tolerance on the spreadsheet and enter 10% as its value. Then click on the Display button to cause the value of the tolerance to be displayed. The edited spreadsheet for the R2 resistor is shown in Fig. 46.

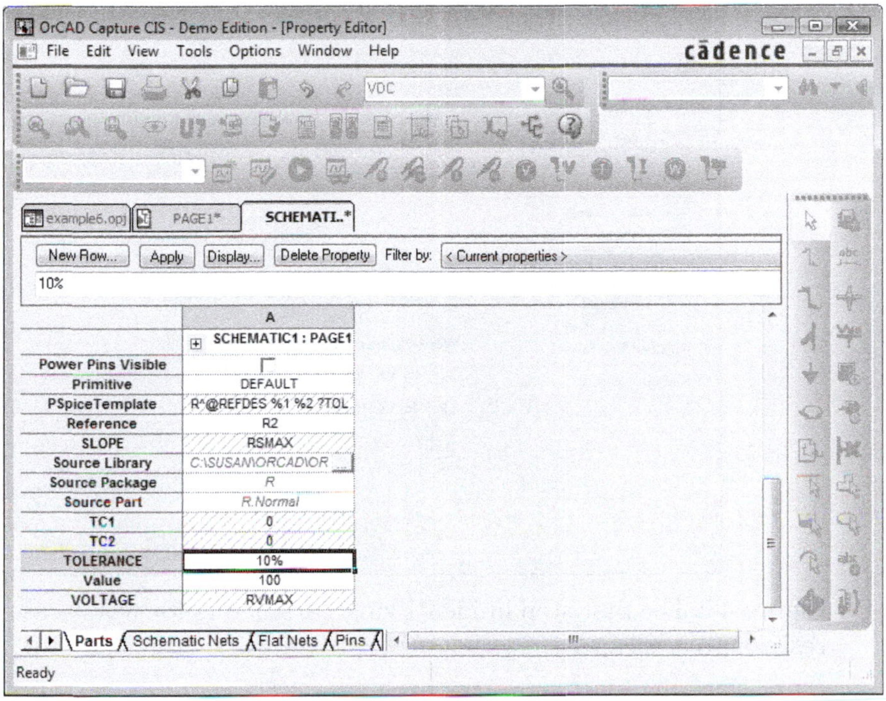

Figure 46: The edited property spreadsheet with a 10% tolerance for the R2 resistor in Fig. 45.

Next, we turn to the analysis selection. We choose DC Sweep analysis, but set both the minimum and maximum values to sweep our input voltage to 125 V, with an increment of 1 V. This screen is not shown. Then we choose the Monte Carlo/Worst Case option, as seen in Fig. 47, select the Worst Case/Sensitivity button, identify the output variable, and choose to vary devices that have device tolerances only. Once these settings have been specified, click on the More Settings button to activate the dialog shown in Fig. 48. To simulate the circuit and examine the effect of resistor tolerance we perform two worst-case analyses—one to determine the maximum worst-case output voltage, and the other to determine the minimum worst-case output voltage. The parameters for the maximum worst-case analysis are shown in Fig. 48, and the relevant portion of the output file is shown in Fig. 49. This figure shows that if the value output resistor, R2, is 10% larger than its nominal value of $100\,\Omega$, the output voltage will be 101.85 V. You can confirm this result using the voltage division equation for the circuit in Fig. 41 when the R2 resistor has the value $110\,\Omega$.

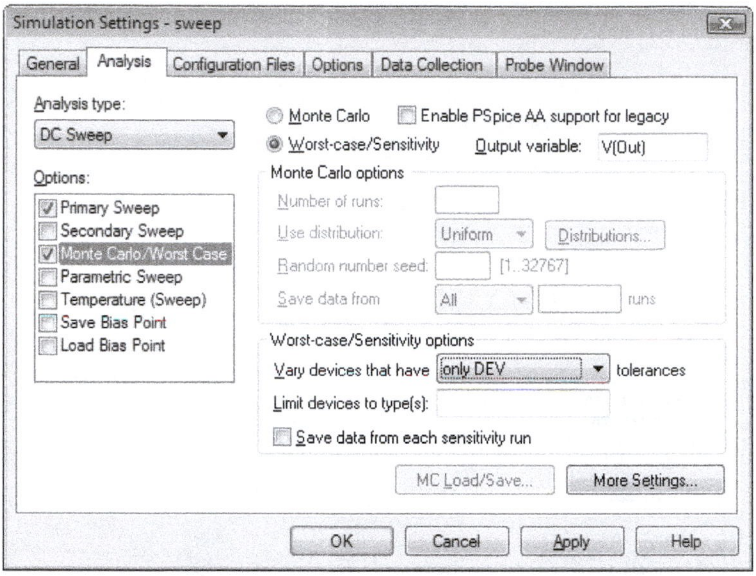

Figure 47: The first step in identifying parameters for worst-case analysis of resistor tolerance.

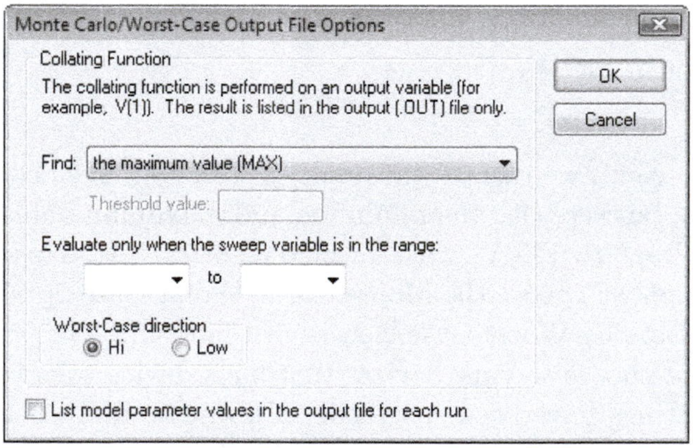

Figure 48: The second step in identifying parameters for worst-case analysis of resistor tolerance.

The second worst-case analysis is specified using the parameters shown in Fig. 50. The results are shown in Fig. 51. Now we see that if the value of the output resistor, R2, is 10% smaller than its nominal value of 100 Ω, the output voltage will be 97.826 V. You can confirm this result using the voltage division equation for the circuit in Fig. 41 when the R2 resistor has the value 90 Ω. Thus, we see that when the R2

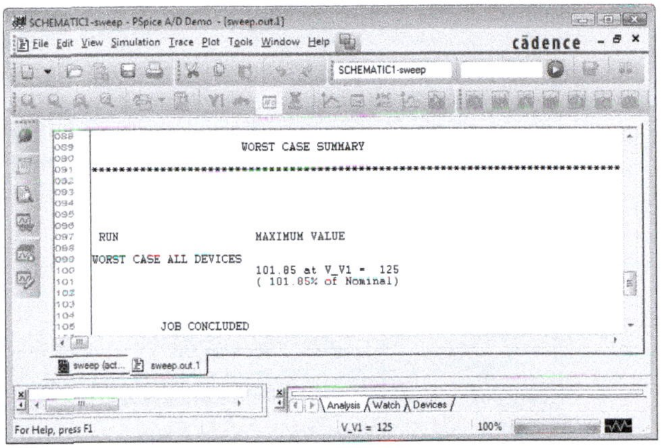

Figure 49: Output of the maximum worst-case analysis of the voltage divider in Fig. 41 using a 10% output resistor.

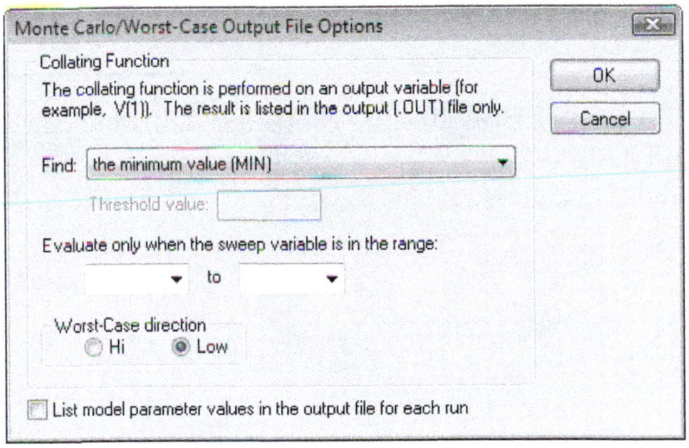

Figure 50: Specifying a worst-case analysis which calculates the minimum output voltage value.

resistor in Fig. 41 has a tolerance of 10%, the actual output voltage may range from 97.826 V up to 101.85 V. If you were designing this circuit in an actual application, you would have to consider whether this variation around the designed value of 100 V is acceptable.

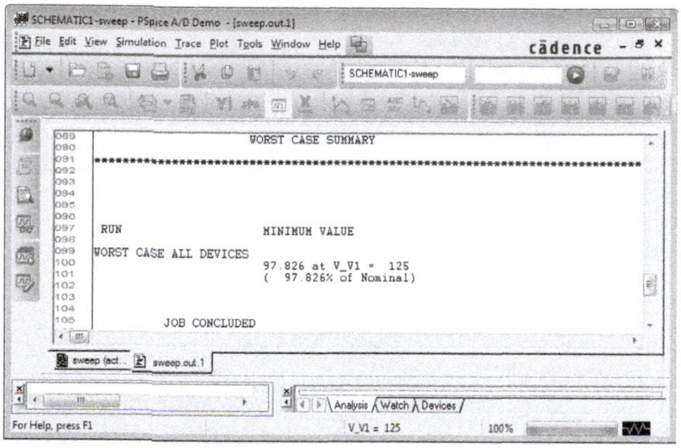

Figure 51: Output of the minimum worst-case analysis of the voltage divider in Fig. 41 using a 10% output resistor.

ASSESSMENT

- Use PSpice to solve the following problems: 4.63, 4.66, 4.67, 4.73, 4.105, 4.106, and 4.107.

Chapter 5

OPERATIONAL AMPLIFIERS

PSpice offers three options for describing an operational amplifier (op amp) within a circuit schematic. The first is to model the op amp with resistors and dependent sources. The second is to take advantage of op amp models already supplied with PSpice. These models are available in a device library. The library models are considerably more sophisticated than the simple dependent-source-based models and are designed to mimic the characteristics of actual op amps. The third is to modify an existing op amp model so that it exhibits the characteristics you desire. This modified model can then be used like any other PSpice circuit element, such as a resistor, and included in multiple locations within the circuit schematic.

In the sections below, the examples illustrate each of these three options for incorporating op amps in your circuit schematics.

5.1 MODELING OP AMPS WITH RESISTORS AND DEPENDENT SOURCES

A simple equivalent circuit for an operational amplifier is given in Fig. 52. Example 7 illustrates how to use PSpice to analyze a circuit containing an operational amplifier and using a simplified equivalent circuit to represent the op amp.

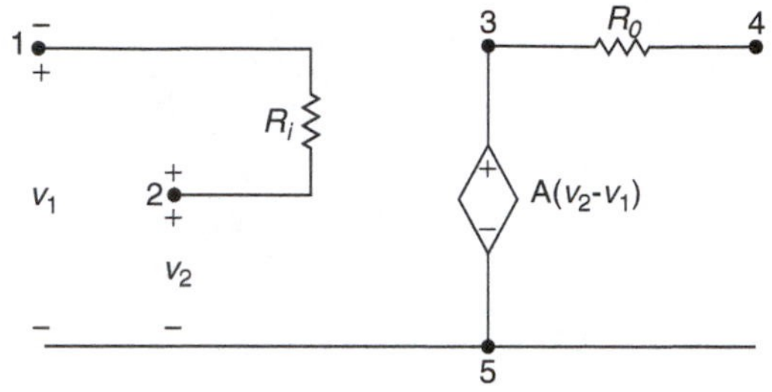

Figure 52: An equivalent circuit for an operational amplifier.

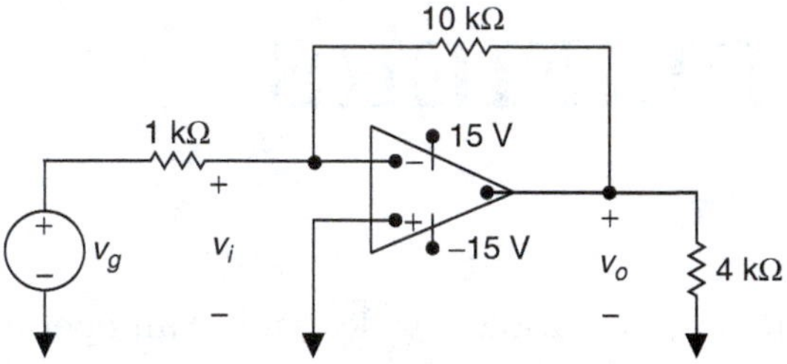

Figure 53: The circuit for Example 7.

EXAMPLE 7

The parameters for the operational amplifier in the circuit shown in Fig. 53 are $R_\text{i} = 200$ kΩ, $A = 10^4$, and $R_\text{o} = 5$ kΩ.

a) Use PSpice to find v_i and v_o when $v_g = 1$ V(dc).

b) Compare the PSpice solution with the analytic solution for v_i and v_o.

SOLUTION

a) The schematic for the circuit in Fig. 53 is given in Fig. 54. We used simple bias point analysis, which produced the output file, a part of which is shown in Fig. 55. Therefore,

$$\begin{aligned} V(\text{IN}) &= 0.0027 \text{ V} = v_i; \\ V(\text{OUT}) &= -9.9697 \text{ V} = v_o. \end{aligned}$$

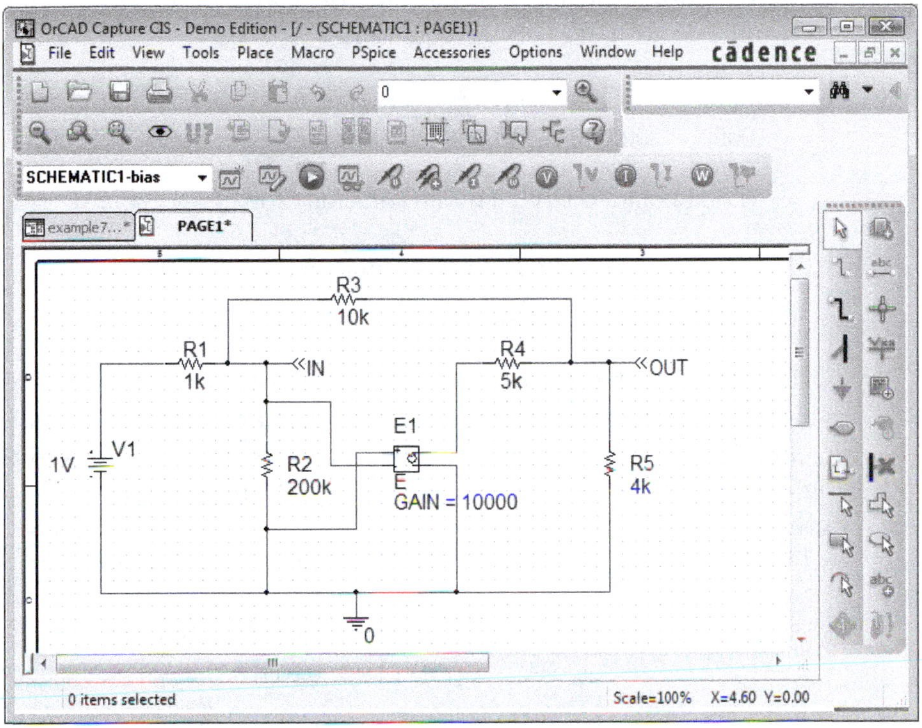

Figure 54: The circuit shown in Fig. 53 represented in schematic form for PSpice analysis.

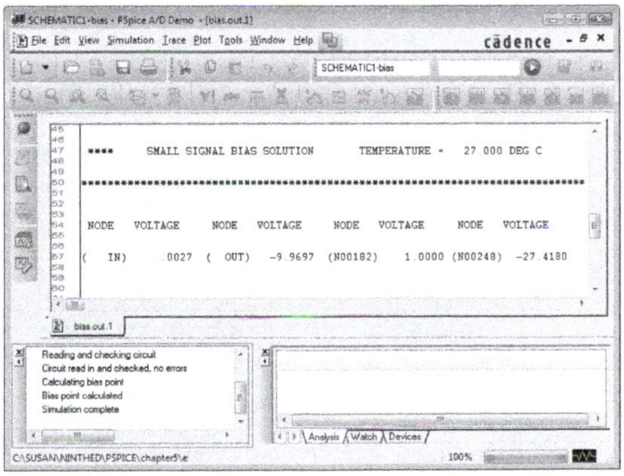

Figure 55: The PSpice output file from the bias point analysis of the circuit in Fig. 54.

b) We obtain the analytic solution for v_i and v_o by solving the following simultaneous node-voltage equations:

$$\frac{v_i - 1}{1000} + \frac{v_i}{200{,}000} + \frac{v_i - v_o}{10{,}000} = 0; \frac{v_o - v_i}{10{,}000} + \frac{v_o + 10^4 v_i}{5000} + \frac{v_o}{4000} = 0.$$

You should verify that the solutions for v_i and v_o are

$$v_i = 2.742 \text{ mV} \quad \text{and} \quad v_o = -9.970 \text{ V}.$$

If we assume that the operational amplifier in the circuit shown in Fig. 53 is ideal, we can alter the PSpice op amp schematic representation by making R_i and A very large and $R_o = 0$. To illustrate, we modify the schematic in Fig. 54, with R2 = 200 MΩ, A = 10^8, and R4 removed from the schematic entirely. The resulting PSpice output file after bias point analysis is shown in Fig. 56. The values of v_i and v_o from this simulation are

$$\begin{aligned} v_i &= \text{V(IN)} = 100\text{E}^{-9} \text{ V} \approx 0 \text{ V}; \\ v_o &= \text{V(OUT)} = -10.0 \text{ V}. \end{aligned}$$

These results are consistent with the analysis of the circuit shown in Fig. 53 if we assume that the operational amplifier is ideal---that is, $v_i = 0$ V and $v_o = -10$ V.

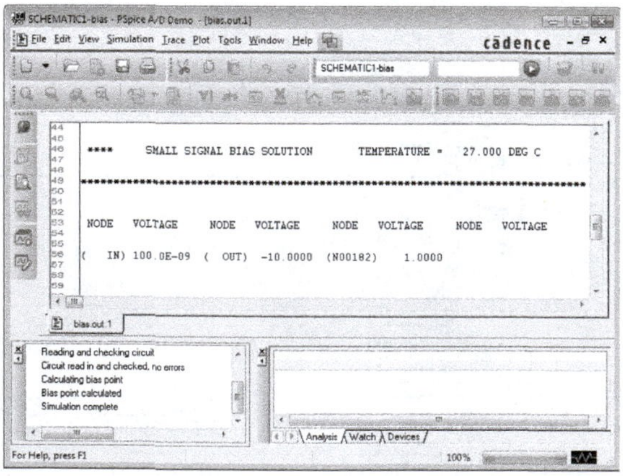

Figure 56: The PSpice output file for the circuit in Fig. 53, for an ideal op amp.

5.2 USING OP AMP LIBRARY MODELS

One of the most powerful features of PSpice is the library of models for various electronic devices. The full version of PSpice has models for more than 5000 devices. The evaluation version of PSpice (used to produce the examples in this supplement) has models for about 100 of the most commonly used devices. Among these models are several for op amps, including the μA741. They are all found in the Eval library. There is also a model for an ideal op amp, which is found in the Analog library. The next two examples repeat Example 7, but incorporate different op amp models in the circuit schematic for Fig. 53.

EXAMPLE 8

Repeat the analysis of the circuit shown in Fig. 53, using the ideal op amp model from the Analog library.

SOLUTION

Figure 57 shows a schematic for the circuit in Fig. 53, where the ideal op amp model is used. Note that this is a simple three-terminal model, with two input terminals for the inverting and non-inverting op amp inputs, and one output terminal for the op amp output. We use bias point analysis to determine the output voltage in response to a 1 V input to this inverting amplifier configuration. The result of this analysis gives an output voltage of -9.9999 V, as seen in Fig. 58, which is essentially the expected output voltage of -10 V.

EXAMPLE 9

Repeat the analysis of the circuit shown in Fig. 53, but this time use the μA741 model from the PSpice model library.

SOLUTION

Figure 59 shows a schematic for the circuit in Fig. 53 that employs the PSpice part that models the μA741 op amp. As you can see, there are 7 different nodes on the op amp model. Nodes 2 and 3 correspond to the inverting and noninverting terminals of the op amp, respectively, and are

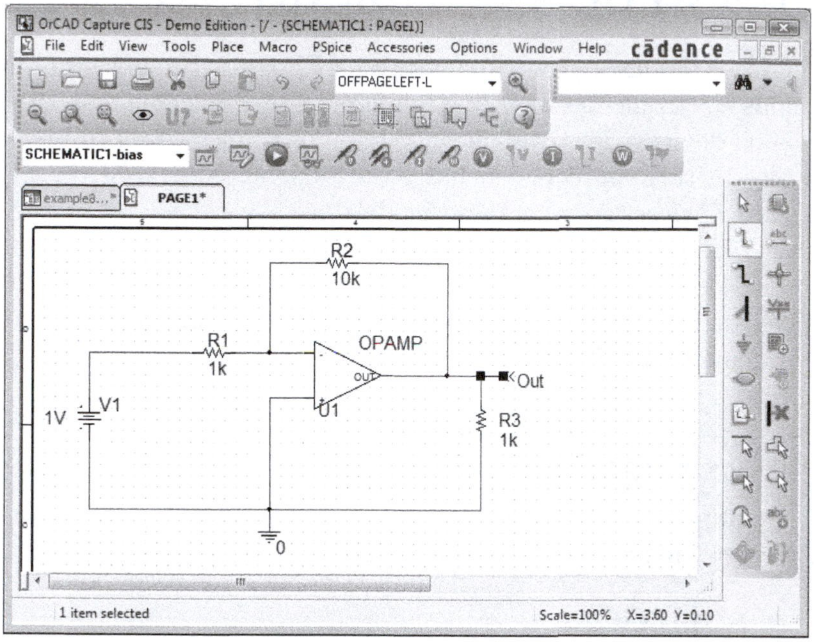

Figure 57: The PSpice schematic for the circuit in Fig. 53, using the PSpice model of an ideal op amp from the Analog library.

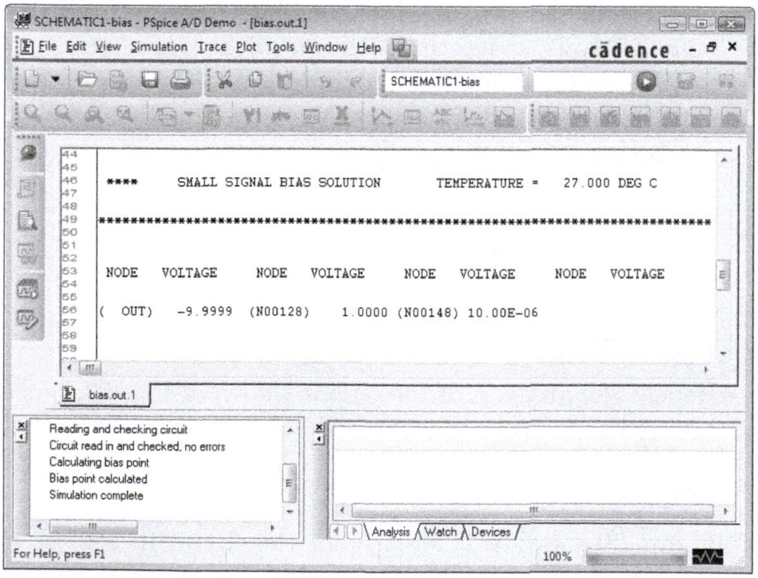

Figure 58: The PSpice output for the bias point analysis of the schematic in Fig. 57.

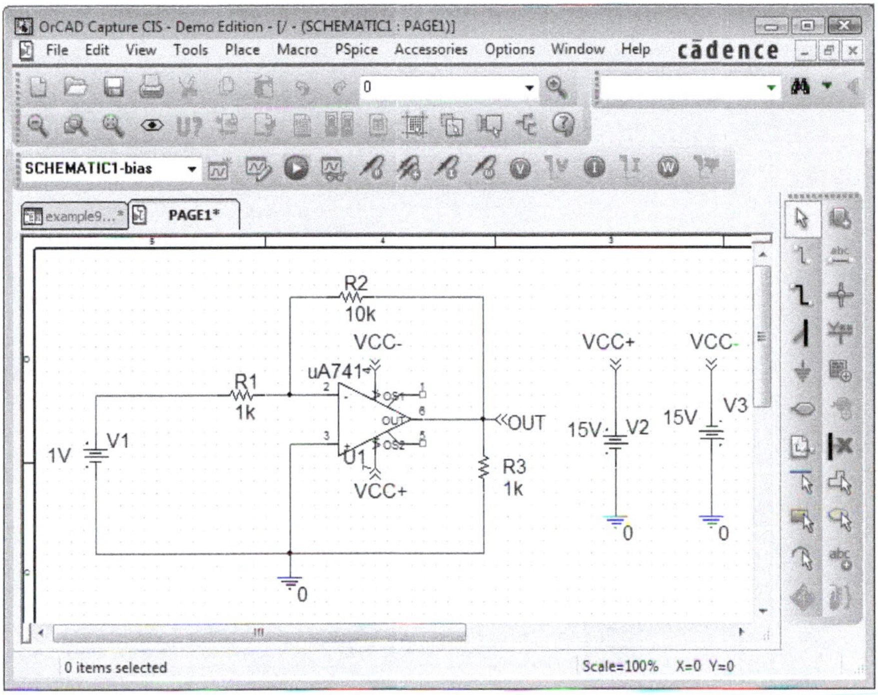

Figure 59: The PSpice schematic for the circuit in Fig. 53, using the PSpice model of the μA741 op amp.

wired in the schematic according to the connections to those terminals in Fig. 53. Node 6 corresponds to the output node of the op amp, and is also wired just like the corresponding node in Fig. 53. Nodes 4 and 7 are the negative and positive power supply connections, respectively. The corresponding nodes in Fig. 53, labeled −15 V and 15 V, are not explicitly wired to any other circuit components. In the schematic, these nodes must actually be connected to dc power supplies. To do this, we use node labels VCC+ and VCC-, and wire these labels to the appropriate dc power supplies to the right of the op amp circuit. This technique for drawing the schematic makes it less cluttered than it would be if we wired the dc power supplies directly into the op amp nodes. Finally, nodes 1 and 5 are not required to be connected to anything, so we leave them as is.

We can use simple bias point analysis to simulate the response of this circuit to its 1 V dc input. The relevant portion of the PSpice output file is shown in Fig. 60. Here, we discover that the output voltage is

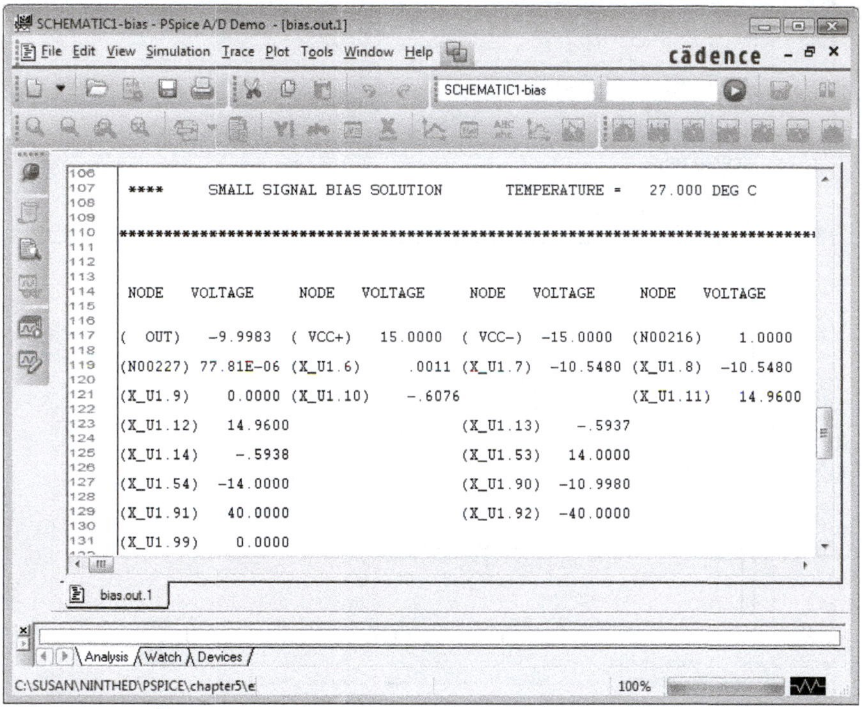

Figure 60: The PSpice output for the bias point analysis of the schematic in Fig. 59.

−9.9983 V, close to the value predicted by the analysis in the previous section.

The model of the circuit for the μA741 component is quite complex. To see the model, highlight the μA741 component in the schematic shown in Fig. 59, right click the mouse and select Edit PSpice Model. The result is shown in Fig. 61. As you can see, the model defines a sub-circuit with over thirty parts, including resistors, capacitors, and models of diodes and transistors. In the full version of PSpice you can actually change the component models to reflect the actual values of the sub-circuit components in a real device you are using. Unfortunately, the Demo Edition that this manual uses does not support the model editor for any PSpice components except a simple diode.

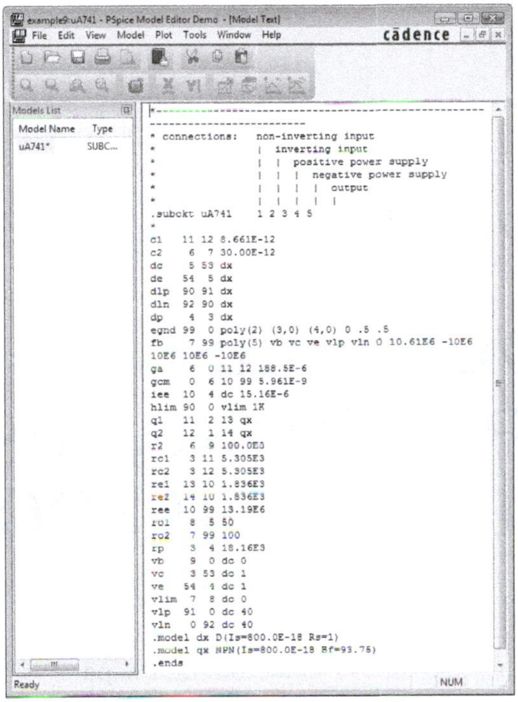

Figure 61: The Model Editor for the uA741 op amp component.

ASSESSMENT

- Use PSpice to solve the following problems: 5.4, 5.6, 5.15(a), 5.18(a), 5.28(a), and 5.39.

Chapter 6

INDUCTORS, CAPACITORS, AND NATURAL RESPONSE

Components that model inductors and capacitors are similar to those used for resistors. The primary property for these components is the value of the inductance (in henries) or the capacitance (in farads). For cases where inductors carry initial current, the initial current can be specified in the Property Editor by modifying the IC property. For cases where capacitors carry initial voltage, the Property Editor is again used to specify the initial voltage, by modifying the IC property.

6.1 TRANSIENT ANALYSIS

You use transient analysis to examine the response of a circuit as a function of time. Selecting Time Response (Transient) in the simulation settings dialog will cause PSpice to simulate the time response of the circuit. You must specify the length of time for the simulation, in seconds, and the time at which data should be stored for output. The duration of the simulation depends on the time constants in the circuit, so some preliminary analysis of the circuit is usually necessary. The start time is almost always zero, unless you want to collect data only after the circuit is in its steady-state.

You have the option to specify the step size, or time increment, to be used by PSpice, which is normally not necessary to specify. If you have used the Property Editor to set initial conditions for any inductors or capacitors in your schematic, you should check the box that will skip the initial transient analysis of the circuit. We'll see these settings in the upcoming example.

You may examine the output of a transient analysis in two different ways. First, you may direct PSpice to write the results of transient analysis to the output file, by clicking the Output File Options button in the simulation settings dialog for time response. This invokes a second dialog, in which you can specify the time increment to use when printing. Second, you may use the program to generate plots of voltage and current versus time.

6.2 NATURAL RESPONSE

To find the natural response of a circuit, you first find the initial inductor currents and capacitor voltages. When you know these values, you can draw the schematic and edit the component properties for transient analysis. Example 10 illustrates how to use PSpice to find the natural response of a series *RLC* circuit. We incorporated a preliminary analytic solution as part of the example to permit checking the validity of the PSpice solution.

EXAMPLE 10

a) Analyze the natural response of the circuit shown in Fig. 62 with respect to the type of damping, the peak value of v_c, and the frequency of oscillation.

b) Use PSpice to analyze the circuit's natural response. Then use the program to generate a plot of the voltage v_c versus t, and use the Probe cursor to identify the peak value of the voltage.

SOLUTION

a) For the series circuit shown in Fig. 62, $\alpha = R/2L = 1000$, $\alpha^2 = 10^6$, and $\omega_o^2 = 1/LC = 50 \times 10^6$. Hence the response is underdamped, and the roots of the characteristic equation are $s_1 = -1000 + j7000$ rad/s and

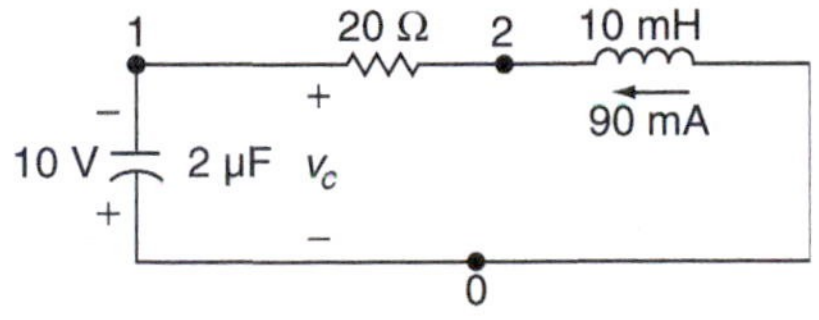

Figure 62: The circuit for Example 10.

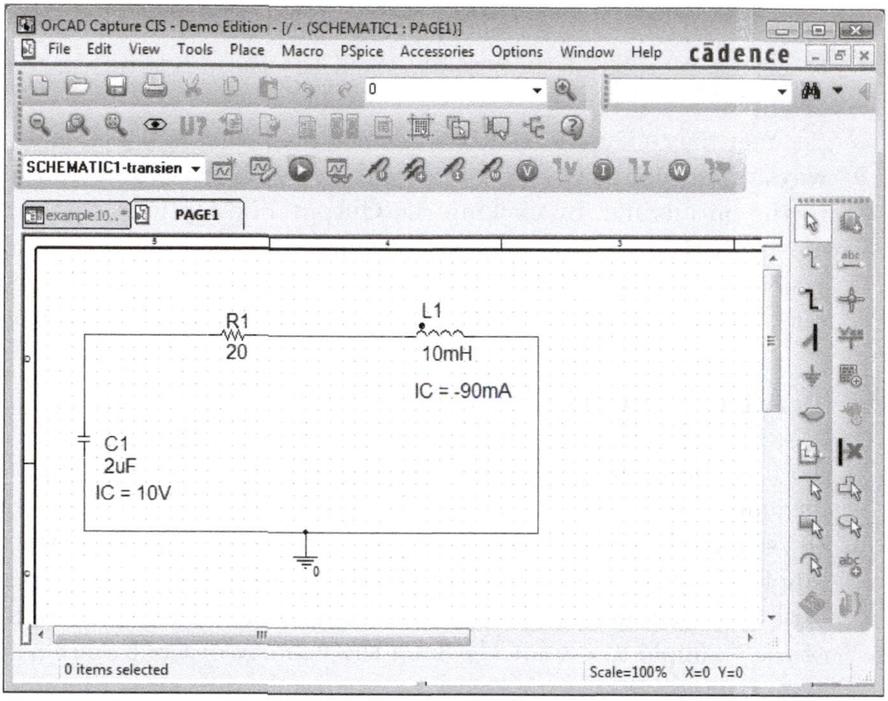

Figure 63: The schematic representation of the circuit in Fig. 62.*

$s_2 = -1000 - j7000$ rad/s. Therefore, the form of the solution for v_c is

$$v_c = (B_1 \cos 7000t + B_2 \sin 7000t)e^{-1000t}.$$

From the initial conditions,

$$v_c(0) = -10 \quad \text{and} \quad \frac{dv_c(0)}{dt} = 45,000 \text{ V/s}.$$

Solving for B_1 and B_2 yields

$$v_c = (-10 \cos 7000t + 5 \sin 7000t)e^{-1000t} \text{ V}, \quad t \geq 0.$$

Based on the solution for v_c, the peak value of v_c is 7.70 V, and it occurs at $t = 362.29$ μs. The damped frequency is 7000 rad/s, and thus the damped period is 897.60 μs.

b) The PSpice schematic representation of the circuit in Fig. 62 is shown in Fig. 63. For both the capacitor and the inductor, we have used the

*In this schematic and any others containing inductors, we have edited the inductor part to eliminate the terminal labels 1 and 2, to simplify the schematic.

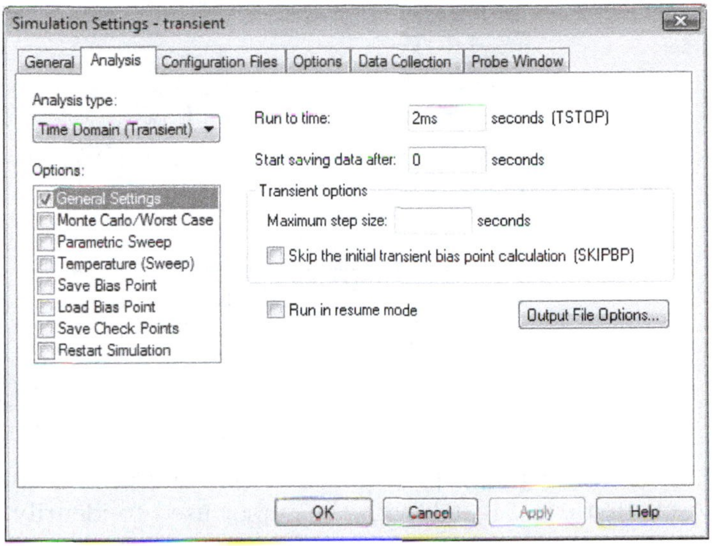

Figure 64: The Simulation Settings for transient analysis of the schematic in Fig. 63.

property editor to set the initial values of voltage for the capacitor and current for the inductor. To get the signs right, you must pay attention to the orientation of these parts. The initial voltage is positive at the left-most terminal of the capacitor, or bottom terminal if the part is rotated once. Therefore, the initial capacitor voltage is 10 V. The initial current enters the left-most terminal of the inductor, or bottom terminal if the part is rotated once. Therefore, the initial inductor current is −90 mA. These values are set in the Property Editor, in the column labeled IC.

From the preliminary analysis of the circuit, we select TSTOP to be 2 ms in the Time Response simulation settings, so that we can analyze the first two cycles of the response. We also check the box that skips the computation of the initial transient bias point. The resulting simulation setting dialog is shown in Fig. 64.

Figure 65 presents the output from the plot of capacitor voltage versus time. You invoke the cursor by selecting Trace/Cursor/Display, and then moving the cursor manually with the right and left arrow keys, or by selecting Trace/Cursor/Peak once the cursor is displayed. We see from the cursor output window that the peak output voltage is 7.7417 V, which occurs at $t = 372.549\,\mu$s, which is in good agreement with our analytic results in (a).

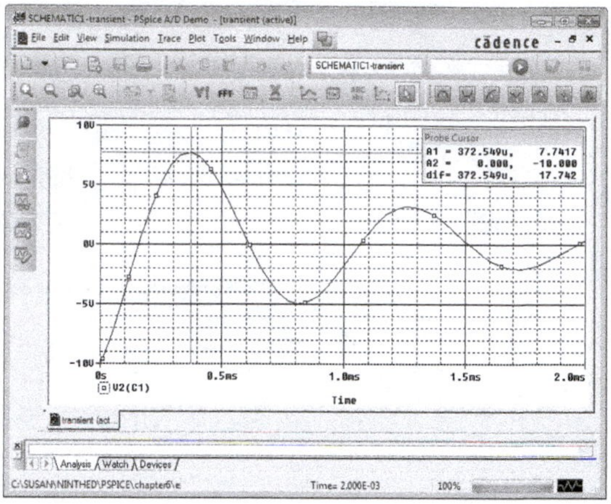

Figure 65: The Probe plot of the output voltage across the capacitor for the schematic in Fig. 63, with the cursor used to identify the peak value of output voltage.

ASSESSMENT

- Use PSpice to find and plot the indicated voltages and currents in the following problems: 8.2, 8.3, 8.4, 8.29, 8.30, and 8.31.

Chapter 7

THE STEP RESPONSE AND SWITCHES

The easiest way to find the step response of a circuit is to insert a dc source at the point where the step change takes place. This is illustrated in Example 11. Then, we present two of several ways to model switches in PSpice. The first involves a specially constructed voltage source, and the second involves insertion of a component that models a realistic switch. We use these two models of switches in Examples 12 and 13.

7.1 SIMPLE STEP RESPONSE

To illustrate the insertion of a dc source to model the step response, we return to the circuit used in Example 10 (Fig. 62) and define the problem in Example 11.

EXAMPLE 11

a) The switch in the circuit shown in Fig. 66 is closed at $t = 0$. At the

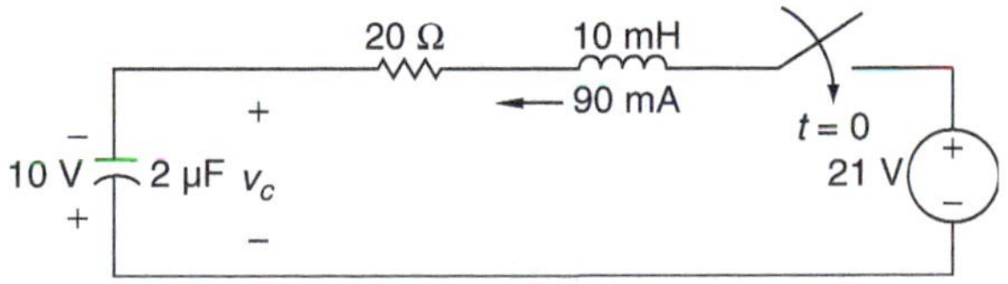

Figure 66: The circuit for Example 11.

instant the switch is closed, the initial current in the 10 mH inductor is 90 mA, and the voltage across the capacitor is 10 V. Figure 66 also shows the references for these initial conditions. Make a preliminary analysis of the step response, and calculate the maximum value of v_c and the time at which it occurs.

b) Create a PSpice schematic, perform transient analysis, and plot v_c versus t from 0 to 3000 μs.

c) Compare the PSpice solution with the preliminary analysis.

SOLUTION

a) From the solution of Example 10, we already know that the response is underdamped. Furthermore, we know that $s_1 = -1000 + j7000$ rad/s, that $s_2 = -1000 - j7000$ rad/s, and that the damped period is 897.60 μs. The step-response solution for v_c takes the form

$$v_c = v_f + (B_1' \cos 7000t + B_2' \sin 7000t)e^{-1000t}.$$

The initial values of v_c and dv_c/dt are the same as in Example 10, and the final value is 21 V. Hence $B_1' = -31$ V and $B_2' = 2$ V, so

$$v_c = 21 - (31 \cos 7000t - 2 \sin 7000t)e^{-1000t} \text{ V}, \qquad t \geq 0.$$

The maximum value of v_c is 41.22 V at 419.32 μs.

b) The PSpice schematic is shown in Fig. 67 and Figure 68 shows the plot generated by Probe. We used two cursors this time, and moved the second cursor by holding down the Shift key while using the left and right arrow keys. The cursors allow you to read the values of v_c(max) and T_d directly from the Probe plot.

c) The output of the PSpice analysis yields

$$\begin{aligned} v_c(\text{max}) &= 41.381 \text{ V}, \quad \text{at } 436.275\ \mu\text{s}; \\ T_d &= 901.925\ \mu\text{s}; \\ v_f &= 21 \text{ V}. \end{aligned}$$

These results agree with the preliminary analysis.

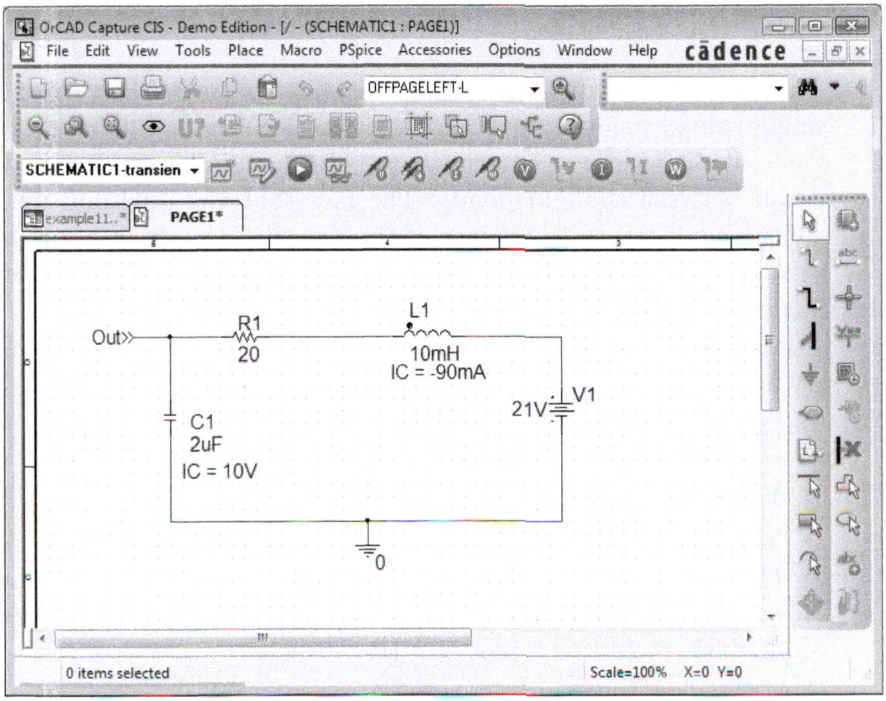

Figure 67: The schematic for the circuit in Fig. 66.

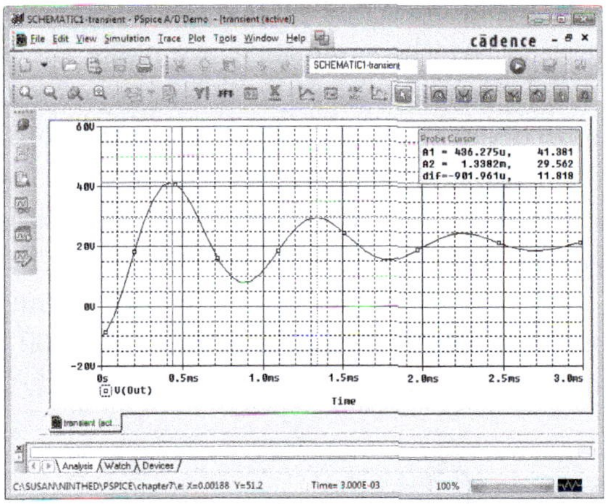

Figure 68: The Probe plot of the results of transient analysis on the schematic in Fig. 67.

7.2 PIECEWISE LINEAR SOURCES

The piecewise linear (PWL) time-dependent source allows you to define a single-valued function at a series of discrete times. PSpice determines the value of the source at intermediate values of time by using linear interpolation between the end points. The fact that the function must be single valued implies that the values of time used in the description of $f(t)$ must continually increase. One of the many uses for this type of source is as a model for a switch.

Example 12 illustrates the use of the piecewise linear source for modeling the switching that takes place in a transient analysis problem.

EXAMPLE 12

The circuit shown in Fig. 69 has been in operation for a long time. At $t = 0$, the 80 V source drops instantaneously to 20 V. Construct a PSpice schematic, using a PWL source, to model this circuit. The results of the analysis should be a plot of $v_o(t)$ versus t.

SOLUTION

To simulate the drop in voltage from 80 to 20 V, we insert a piecewise linear voltage source of opposite polarity in series with the 80 V dc source. We then step the piecewise source from 0 to 60 V over a short time interval. At the same time, we make the duration of the 60 V step long enough that the circuit reaches its new steady-state condition. We must now decide how rapidly the source must rise to 60 V and how long it should remain at 60 V. We leave to you verification that the time constant for the circuit shown in Fig. 69 is 40 ms. We select a rise time for the piecewise linear voltage source of 40 ms/1000, or 40 μs. The duration of the 60 V source should be much longer than 40 ms, so we arbitrarily select 10 time constants, or 400 ms. The piecewise linear voltage source has the waveform shown in Fig. 70.

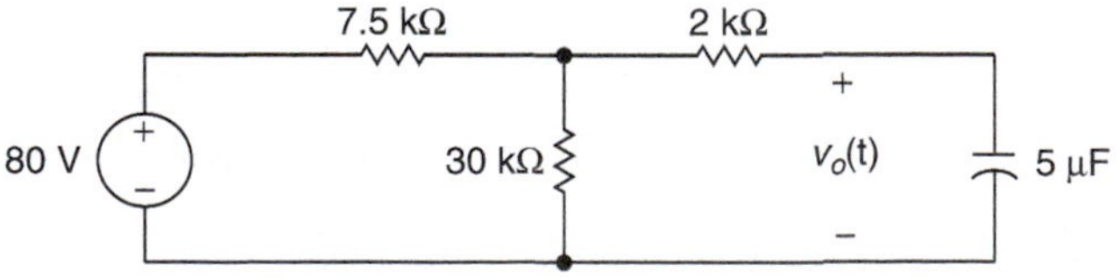

Figure 69: The circuit for Example 12.

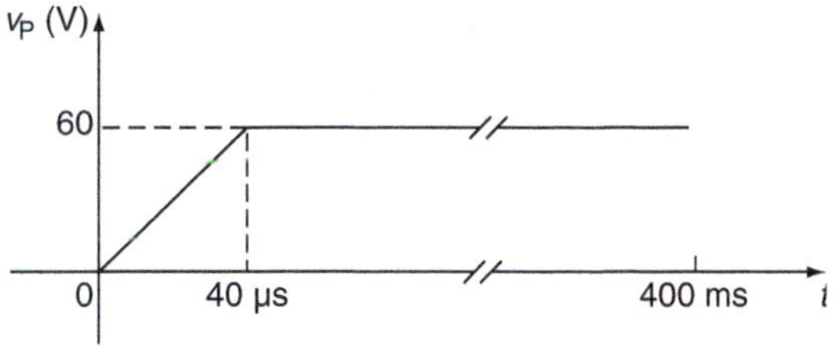

Figure 70: The waveform for the piecewise linear voltage source.

Figure 71 is the circuit schematic for the circuit shown in Fig. 69. We specify the behavior of the piecewise linear voltage source by entering pairs of time and voltage values into the appropriate columns in the Property Editor spreadsheet for this component. The first pair of values, with column headings T1 and V1, are both zero, since the value of the voltage source is 0 V at $t = 0$ s. The next pair of values, with column headings T2 and V2, are 40 μs and 60 V, since the value of the voltage source is 60 V at $t = 40\,\mu$s. The final pair of values, with column headings T3 and V3, are 40 ms and 60 V, since the voltage source continues to supply 60 V for all time after $t = 40\,\mu$s. Thus, the value for T3 is

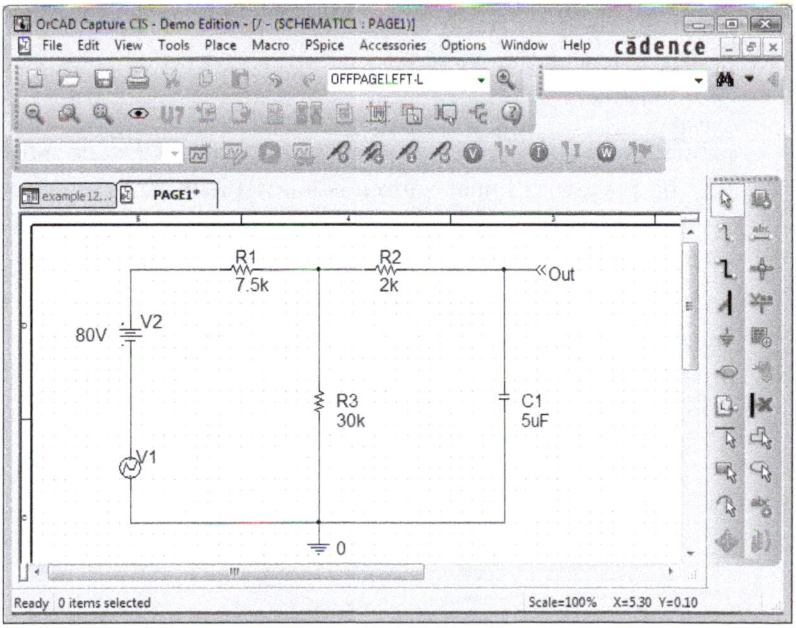

Figure 71: The schematic for the circuit in Fig. 69.

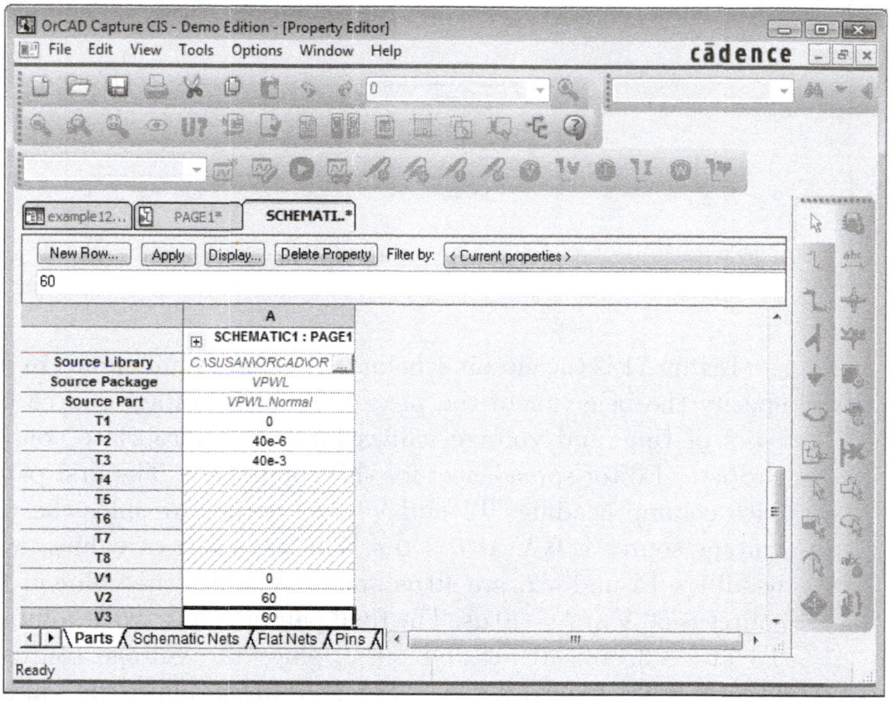

Figure 72: The Property Editor spreadsheet for the piecewise linear voltage source from the schematic in Fig. 71. (Right click on the spreadsheet and select Pivot to get this view.)

arbitrary, so long as it is a time larger than T2. The edited spreadsheet for the piecewise linear source is shown in Fig. 72.

In this type of transient analysis problem, we use PSpice to calculate initial conditions automatically, since there were no prespecified initial conditions. Thus we don't check the SKIPBP box in the simulation settings dialog. The plot in Fig. 73, which uses the cursors to determine the output voltage at $t = 40$ ms and $t = 80$ ms, yields the following results:

t, ms	$v_o(t)$, V
0	64
40	33.654
80	22.489

Furthermore, you can see from the plot that $v_o(t)$ approaches 16 V as t approaches ∞.

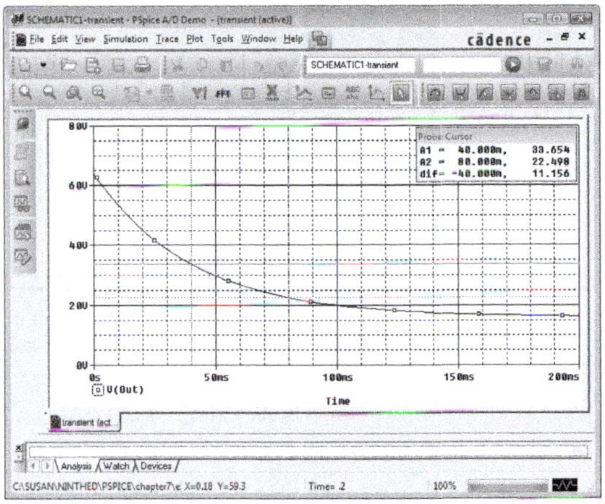

Figure 73: Probe plot of the transient response of the schematic in Fig. 71.

7.3 REALISTIC SWITCHES

PSpice provides two special circuit components that you may use to represent actual switches more accurately. The components are both in the Eval library. One is a normally-opened switch, with the part name `Sw_tOpen` and the other is a normally-closed switch, with the part name `Sw_tClose`. These switches include attributes of a realistic switch, including the equivalent resistance of the switch when open, the equivalent resistance of the switch when closed, the time when the switch opens (if it is a normally-closed switch) or closes (if it is a normally-opened switch), and the transition time for switching from one state to the other.

We illustrate the use of the normally-opened switch in Example 13 to examine the effect of nonzero switch resistance on the step response of the *RLC* circuit.

EXAMPLE 13

Use PSpice to model the circuit shown in Fig. 66, with some modifications. Assume that the voltage source is the same as that used in Example 11, and use the same values in specifying transient analysis. But assume that the resistance of the switch in the ON position is 10 Ω. Then generate a plot of v_{c} versus t. Compare the results here with those in Example 11.

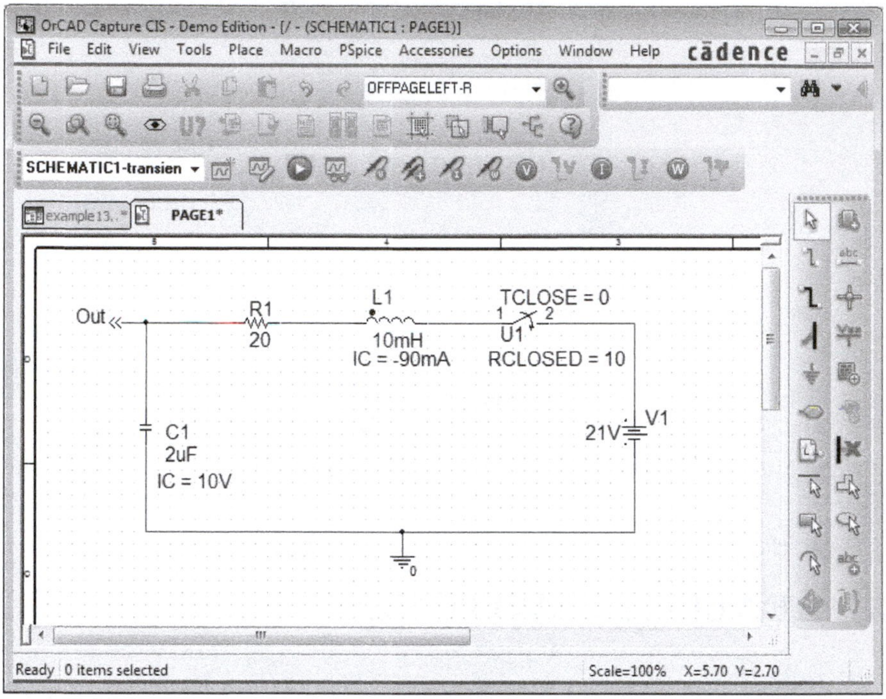

Figure 74: The schematic for the circuit in Fig. 66, using a normally-opened switch.

SOLUTION

The PSpice schematic that models the circuit in Fig. 66 is shown in Fig. 74. We set the attribute values for the switch so that when the switch is closed, its resistance is 10 Ω. The spreadsheet for the switch after this attribute has been modified is shown in Fig. 75. From this figure you can see the other properties of the switch that can be specified. We have displayed the initial condition for the inductor and capacitor by highlighting the IC column in the spreadsheet and clicking the Display button.

Figure 76 shows the plot of v_c versus t. Note that the values of $v_c(\max)$ and T_d are different in this example, as compared to Example 11, because of the additional resistance introduced by the model of the switch.

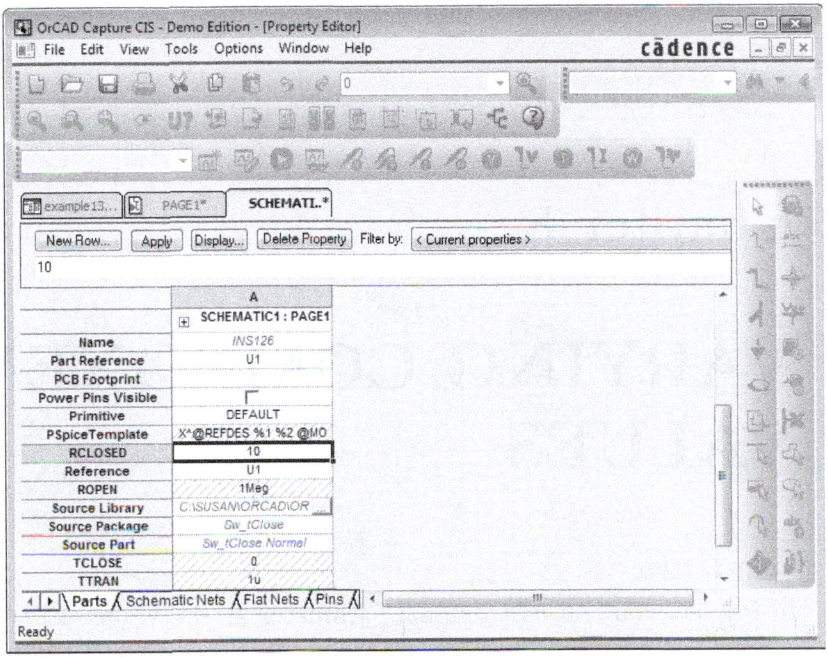

Figure 75: The property spreadsheet for the normally-opened switch in the schematic of Fig. 74.

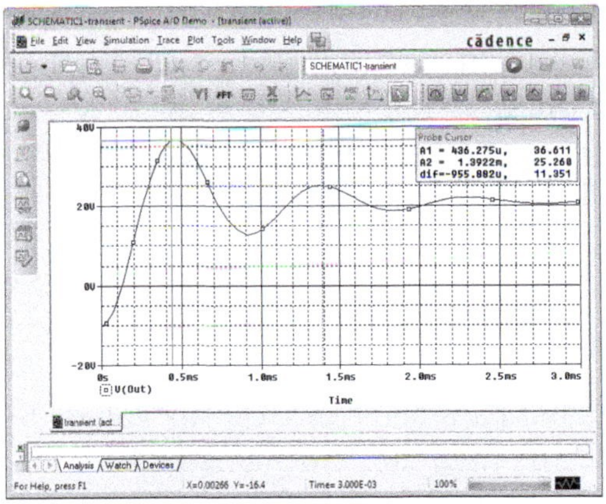

Figure 76: The plot of capacitor voltage for the schematic in Fig. 74.

ASSESSMENT

- Use PSpice to find and plot the indicated voltages and currents in the following problems: 7.12, 7.16, 7.26(a) and (b), 7.30, 7.46, 7.63, 8.36, 8.49, and 8.63.

Chapter 8

VARYING COMPONENT VALUES

In PSpice, you can use a global parameter as the value of a circuit component. Then when you vary the value of the global parameter, the effect is to vary the value of the circuit component. This feature of computer-aided circuit analysis is very powerful, for it allows you to see the effects of changing a circuit parameter on the behavior of the circuit.

Example 14 demonstrates how easily PSpice can analyze the same circuit several times with a different value of a given component used in each analysis. We use Probe to illustrate graphically the varying behavior of the circuit as the component value changes.

EXAMPLE 14

Use a global parameter to vary the value of the resistor in Fig. 63 (see Example 10) from 20 Ω to 100 Ω in 20 Ω steps. Then plot the value of v_C versus t for each of the resistor values.

SOLUTION

The schematic for the circuit in Fig. 63 is shown in Fig. 77. While the capacitor and the inductor are modeled just as they were in Example 10, the model of the resistor is different. Instead of the resistor having a numeric value for its resistance, it has the name of a global parameter,

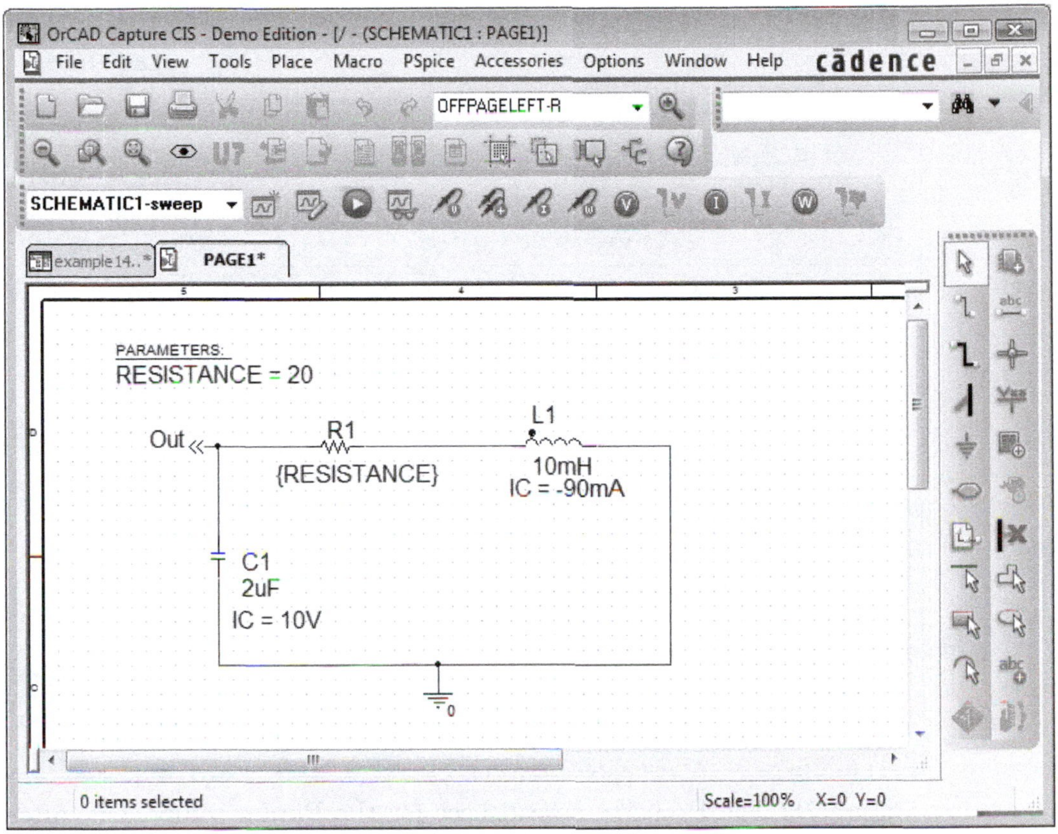

Figure 77: The schematic for the circuit in Fig. 62 with a variable resistance.

in this case {RESISTANCE}. We define the global parameter by adding the Param part from the Special library, as can be seen in Fig. 77.

To define the global parameter, invoke the property editor on the Param component. Then, click the New button, and type the name of the global parameter in the dialog box. This creates a new property, with the name RESISTANCE, for the Param component. We then entered a default value of 20 Ω for this new property. To have this property's name and its value appear in the schematic, highlight the spreadsheet column with the new property and click the display button. Then choose Display Name and Value from the Display Format dialog. The spreadsheet for the Param part, updated as just described, is shown in Fig. 78.

Now that the schematic has completely described the circuit, we turn to specifying the analysis. We again choose Time Domain, or transient,

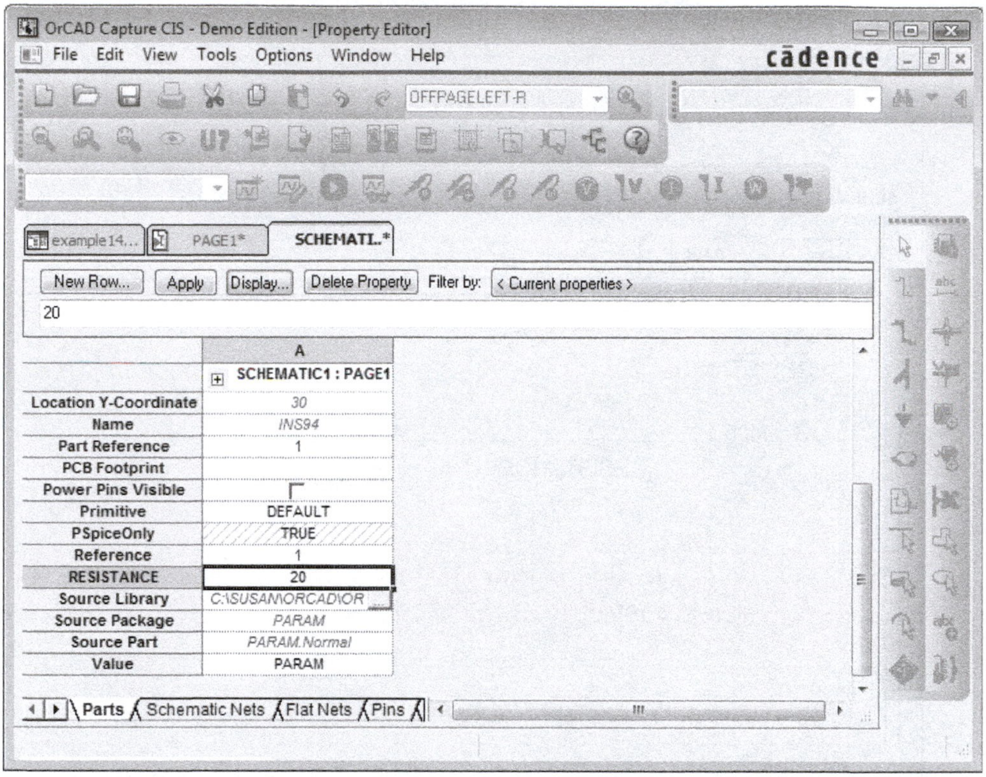

Figure 78: The Property spreadsheet for the Param component, after the RESISTANCE property has been added. (Right-click in the spreadsheet and select Pivot to see this display format.)

analysis, with the same final time as we have used previously with this circuit, namely 2 ms. But now we also check the box labeled Parametric Sweep. As shown in Fig. 79, we then select a Global parameter as the sweep variable, identify the parameter name, and input the start value, end value and increment for the sweep variable.

When we run the transient analysis, PSpice actually simulates transient analysis for five different circuits, one circuit for each of the five different resistance values we specified. We selected all five of these results to plot, as shown in Fig. 80. Note that the analysis using the smallest value of resistance has the smallest damping ratio. Increasing the value of resistance increases the damping ratio. You should confirm these results analytically.

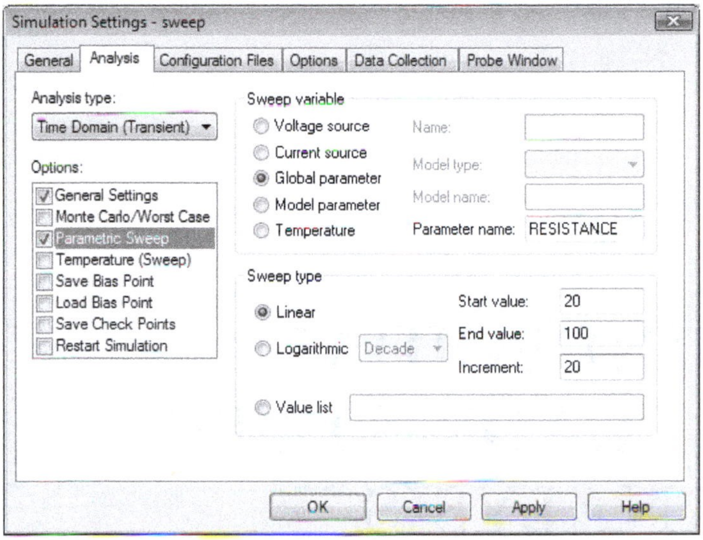

Figure 79: Specifying a parametric sweep of the RESISTANCE global parameter for the schematic in Fig. 77.

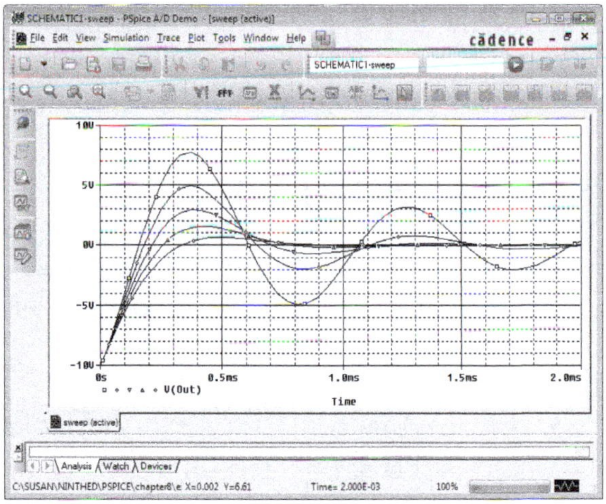

Figure 80: Plot of capacitor voltage versus time as the resistance is varied in the series RLC circuit of Fig. 62. The 20 Ω resistance produces the plot with the highest peak.

ASSESSMENT

- Use PSpice to solve the following problems: 4.68, 4.83, 4.88, 8.18, 8.19, 8.20, and 8.44.

Chapter 9

SINUSOIDAL STEADY-STATE ANALYSIS

We now discuss the features of PSpice that permit you to examine how a circuit behaves in response to a sinusoidal input at a fixed frequency. This behavior is the ac steady-state response of the circuit.

9.1 SINUSOIDAL SOURCES

The following equations describe the damped sinusoidal voltage source, used in simulating a circuit's time-domain response to such an input:

$$
\begin{aligned}
v_{\mathrm{g}} &= \mathrm{VOFF} + \mathrm{VAMPL}\sin[\mathrm{PHASE}], \quad 0 \le t \le \mathrm{TD}; \\
v_{\mathrm{g}} &= \mathrm{VOFF} + \mathrm{VAMPL}e^{-\mathrm{DF}(t-\mathrm{TD})}\sin[2\pi\mathrm{FREQ}(t-\mathrm{TD}) + \mathrm{PHASE}], \\
&\qquad \mathrm{TD} \le t \le \mathrm{TSTOP};
\end{aligned}
$$

where

$$
\begin{aligned}
\mathrm{VOFF} &= \text{offset voltage in volts;} \\
\mathrm{VAMPL} &= \text{amplitude in volts;} \\
\mathrm{FREQ} &= \text{frequency in hertz;} \\
\mathrm{TD} &= \text{delay in seconds;} \\
\mathrm{DF} &= \text{damping factor in seconds}^{-1}\text{; and} \\
\mathrm{PHASE} &= \text{phase angle in degrees.}
\end{aligned}
$$

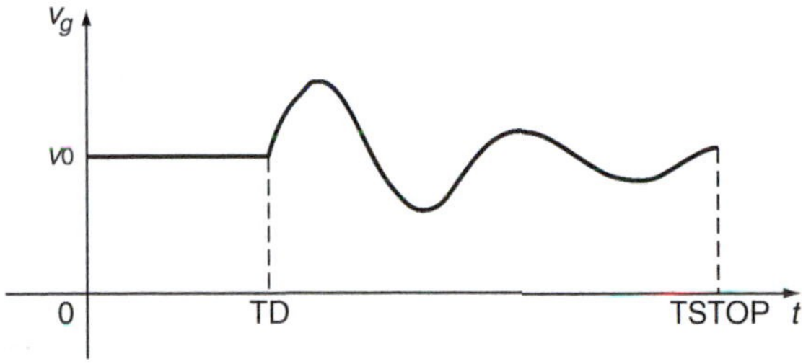

Figure 81: A damped sinusoidal voltage.

The default values of TD and DF are zero, whereas FREQ defaults to 1/TSTOP. Figure 81 shows a graph of the damped sinusoidal voltage.

When constructing any of the time-dependent sources, you should check a plot of its waveform before using it in a specific problem. You can check the waveform by first applying it to a purely resistive circuit. The output waveform in a resistive circuit is a scaled replica of the input waveform. You may always choose a convenient scaling factor for the test circuit. For example, assume that you want to check the data statement for the damped sinusoidal voltage:

$$v_g = 12.5 + 25e^{-0.1(t-1)} \sin[0.4\pi(t-1)] \text{ V}.$$

If you apply this voltage to the voltage-divider circuit shown in Fig. 82, the output voltage is $0.8v_g$. The part name for the sinusoidal voltage source is Vsin and is found in the Source library. Again we turn to the Property Editor to access the property spreadsheet for this part and specify the values for TD, DF, VOFF, VAMPL, FREQ, and PHASE. The edited spreadsheet is shown in Fig. 83. The resulting output voltage for the transient analysis of Fig. 82 is shown in Fig. 84, and confirms that we have correctly specified the properties of the sinusoidal voltage source.

9.2 SINUSOIDAL STEADY-STATE RESPONSE

The sinusoidal sources discussed in the previous section are used when a circuit's time response is of interest. Thus, these sources are intended to be used in conjunction with transient analysis. If we want to analyze a circuit to determine its sinusoidal steady-state response (also known as the ac steady-state response), we use ac voltage or current sources, whose part names are Vac and Iac, respectively. These sources are used together with AC Sweep analysis to calculate the output voltage or current phasors for one frequency

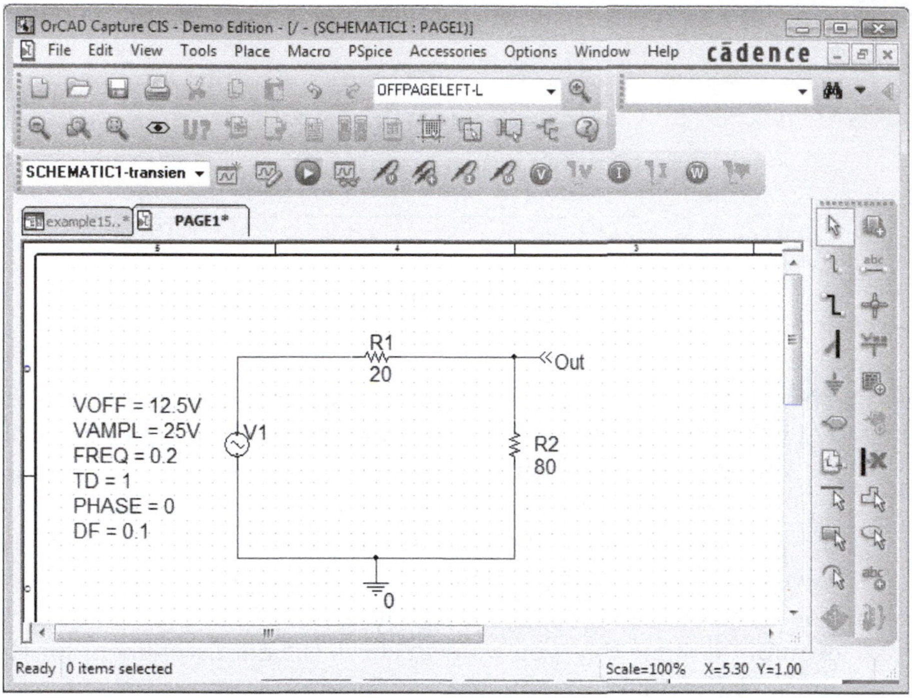

Figure 82: The schematic for a voltage divider used to check a sinusoidal source.

or over a range of different frequency values. Example 15 illustrates the use of ac sources and the AC Sweep analysis to find the single-frequency steady-state sinusoidal response.

EXAMPLE 15

The sinusoidal sources in the circuit shown in Fig. 85 are described by the equations

$$v_g = 20\cos(10^5 t + 90^\circ)\ \text{V} \quad \text{and} \quad i_g = 6\cos 10^5 t\ \text{A}.$$

a) Use PSpice to find the magnitude and phase of v_o and i_o.

b) Check the PSpice results obtained in (a) against an analytic solution for v_o and i_o.

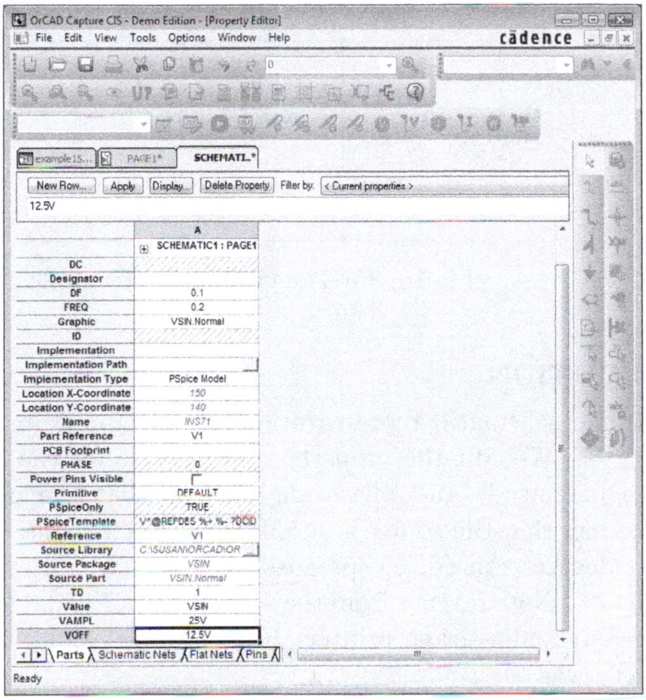

Figure 83: The property spreadsheet for the sinusoidal voltage source in Fig. 82. (Right-click in the spreadsheet and select Pivot to see this display format.)

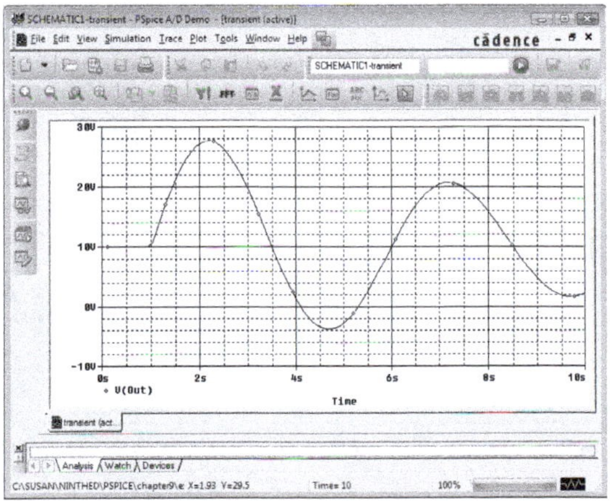

Figure 84: The plot of the output voltage for the transient analysis of the schematic in Fig. 82.

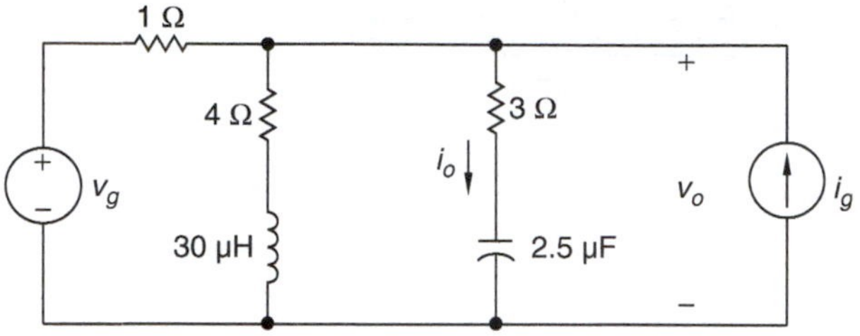

Figure 85: The circuit for Example 15.

SOLUTION

a) The schematic representation of the circuit in Fig. 85 is shown in Fig. 86. We edit the property spreadsheets for the ac sources to specify the magnitude and phase angle. Note that the phase angle property assumes that the source is specified as a cosine, and the phase angle units are degrees. An edited spreadsheet for the ac voltage source is shown in Fig. 87. Note further from the schematic in Fig. 86 that we have inserted current and voltage printers to cause the results of the analysis to be

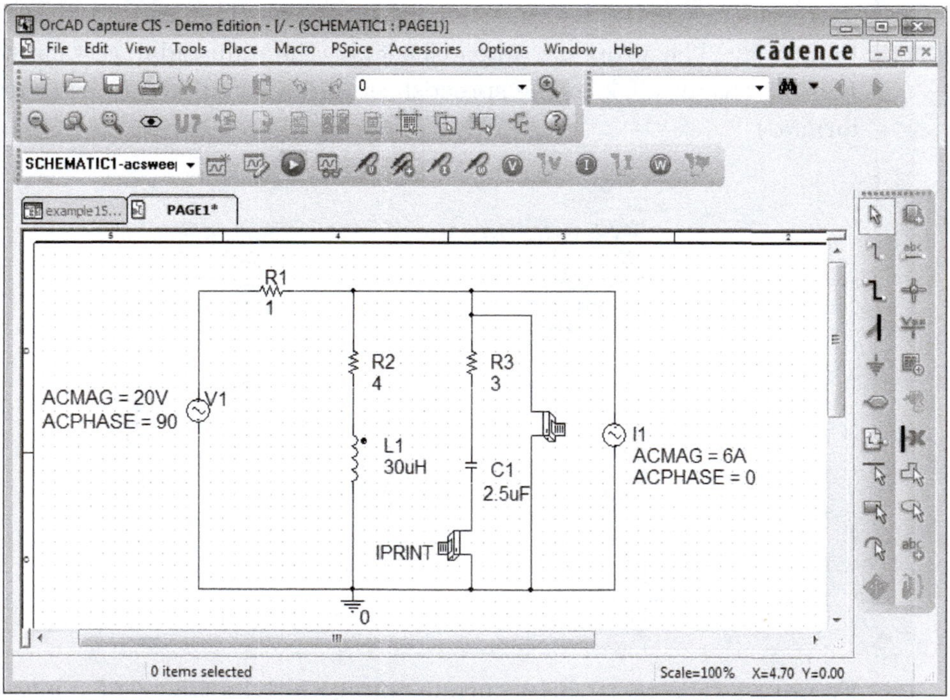

Figure 86: The schematic for the circuit in Fig. 85.

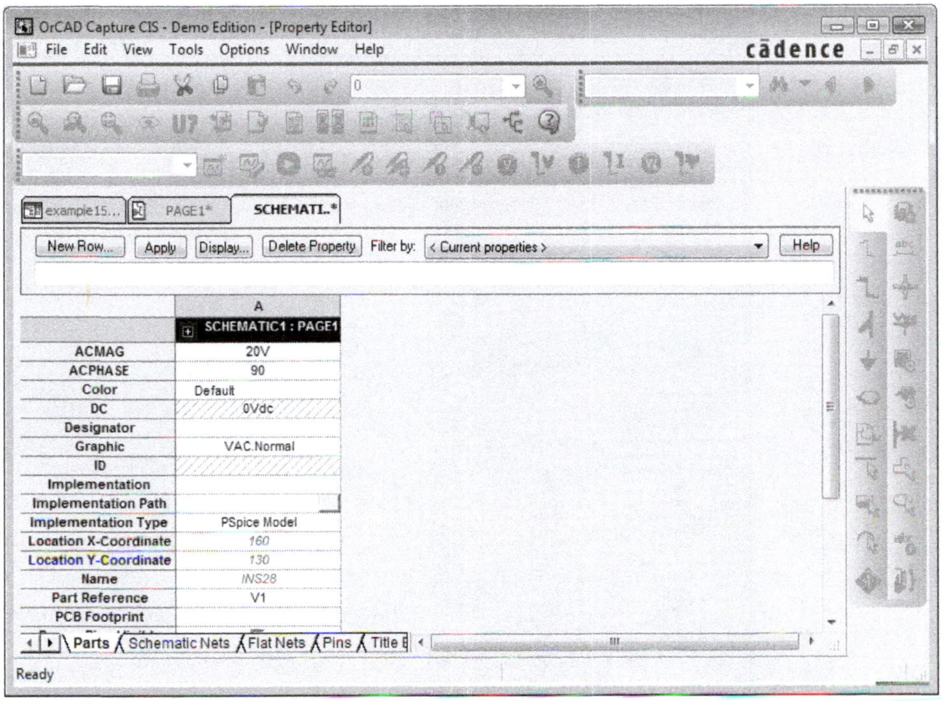

Figure 87: The edited property spreadsheet for the ac voltage source in the schematic of Fig. 86.

printed in the output file. In order to get the desired output data, we must edit the property spreadsheets for these parts. Remember to specify the output from AC analysis by placing a "Y" (for "yes") in the column labeled AC. Place a Y in the columns labeled MAG and PHASE if you want the output values in polar form. Place a Y in the columns labeled REAL and IMAG if you want the output values in rectangular form. The edited property spreadsheet for the voltage printer is shown in Fig. 88. As you can see, we have specified that both the polar and rectangular forms of the output voltage be written to the output file.

Next we specify the type of analysis, which in this case is AC Sweep. The dialog box for this analysis is shown in Fig. 89. We are asked to specify the start and end frequency and the total number of frequency points in our sweep. Since we want PSpice to analyze the schematic for only a single frequency, the start and end frequencies are the same, and the total number of frequency values is 1. Note that we must specify the frequency in a PSpice source file in hertz. The pertinent output from the

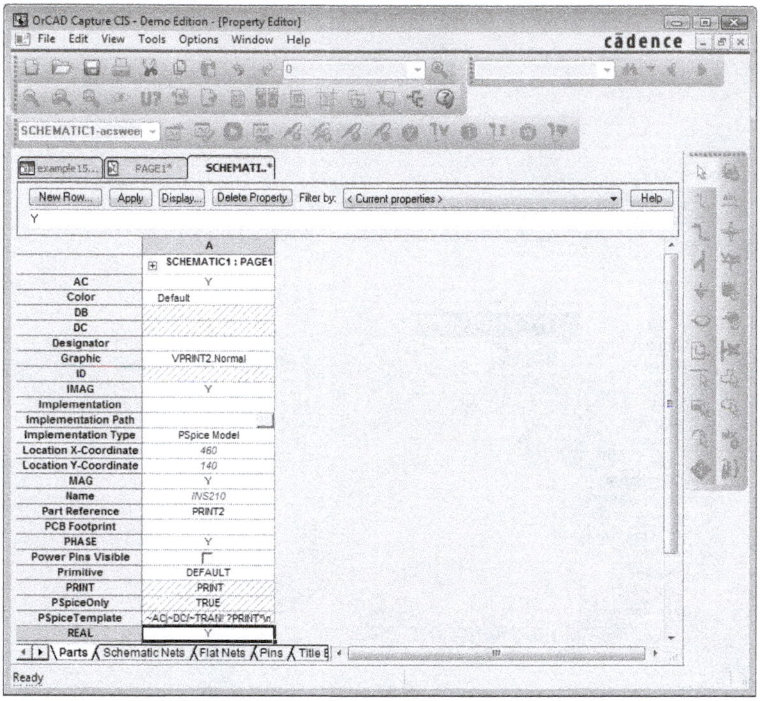

Figure 88: The edited property spreadsheet for the voltage printer in the schematic of Fig. 86.

PSpice analysis is shown in Fig. 90. The PSpice simulation results are $\mathbf{V}_o = 16.31\underline{/71.51^\circ}$ V and $\mathbf{I}_o = 3.261\underline{/124.6^\circ}$ A.

b) We obtain the solution for the phasor voltage $\mathbf{V}_o$ by solving a single node-voltage equation:

$$\frac{\mathbf{V}_o}{4+j3} + \frac{\mathbf{V}_o}{3-j4} + \frac{\mathbf{V}_o - j20}{1} = 6\underline{/0^\circ}.$$

The solution for $\mathbf{V}_o$ is

$$\mathbf{V}_o = 16.31\underline{/71.51^\circ}\text{ V}.$$

Hence

$$\mathbf{I}_o = \frac{\mathbf{V}_o}{3-j4} = 3.26\underline{/124.64^\circ}\text{ A}.$$

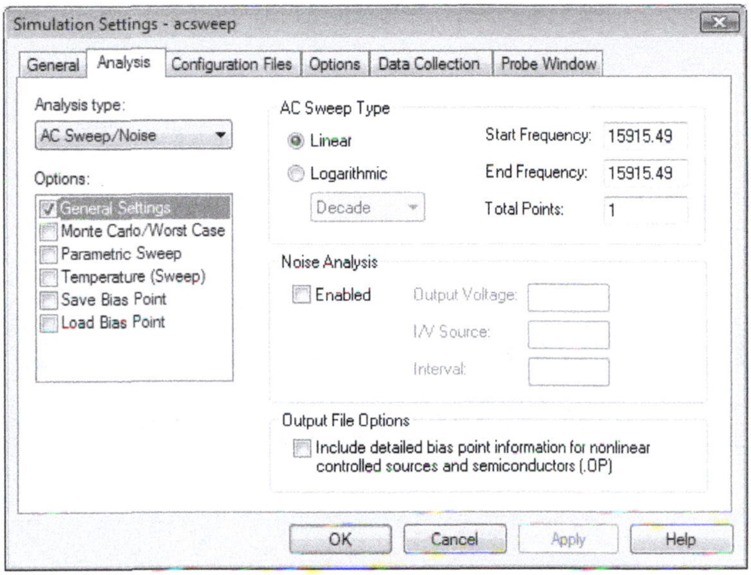

Figure 89: The dialog box for AC Sweep analysis of the circuit in Fig. 86.

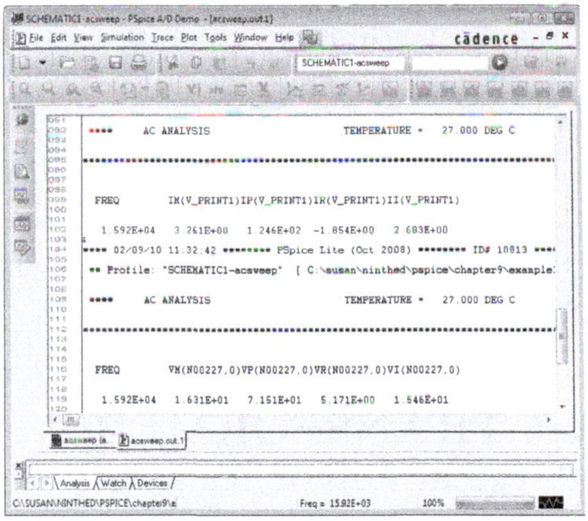

Figure 90: The output from the ac analysis of the schematic in Fig. 86.

ASSESSMENT

- Use PSpice to solve the following problems: 9.29, 9.34, 9.40, 9.54, 9.69, and 9.70.

Chapter 10

LINEAR AND IDEAL TRANSFORMERS

PSpice supports analysis of circuits with three types of mutually-coupled coils. First, it is possible to specify pairs of inductors which are coupled by using simple inductors, just as we have previously, together with a part named K in the Analog library, which is used to select the coupling coefficient for the inductors. You are encouraged to explore PSpice analysis of a circuit with mutually-coupled coils. Both linear and ideal transformers are modeled with the `XFRM_Linear` part from the Analog library. The properties specified for this part include the inductance of the two coils and the coefficient of coupling for the transformer. The linear and ideal transformers are illustrated in the following sections.

10.1 LINEAR TRANSFORMERS

The PSpice part used to model a linear transformer has the name `XFRM_Linear` and is found in the Analog library. We will illustrate the use of this part in Example 16.

EXAMPLE 16

a) Use PSpice to find the rms amplitude and the phase angles of i_1 and i_2 in the circuit shown in Fig. 91 when $v_g = 141.42 \cos 10t$ V.

b) Check the PSpice solution by using it to show that the average power generated equals the average power dissipated.

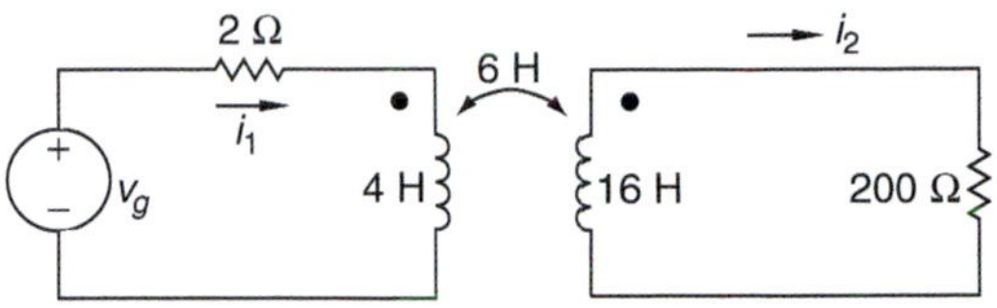

Figure 91: The circuit for Example 16.

SOLUTION

a) Before redrawing the circuit for analysis by PSpice, we note that $V_g = 100\underline{/0^\circ}$ V rms, f= $5/\pi \approx 1.59155$ Hz, and the coefficient of coupling $k = 6/\sqrt{64} = 0.75$. Figure 92 shows the schematic for the circuit. Note that both the primary and the secondary sides of the linear transformer must be explicitly connected to the zero node, or ground. Also, we have inserted current printers to direct the result of the AC Sweep analysis to the output file.

We edit the property spreadsheet to specify the values of the inductance for the primary and secondary coils as well as the coupling

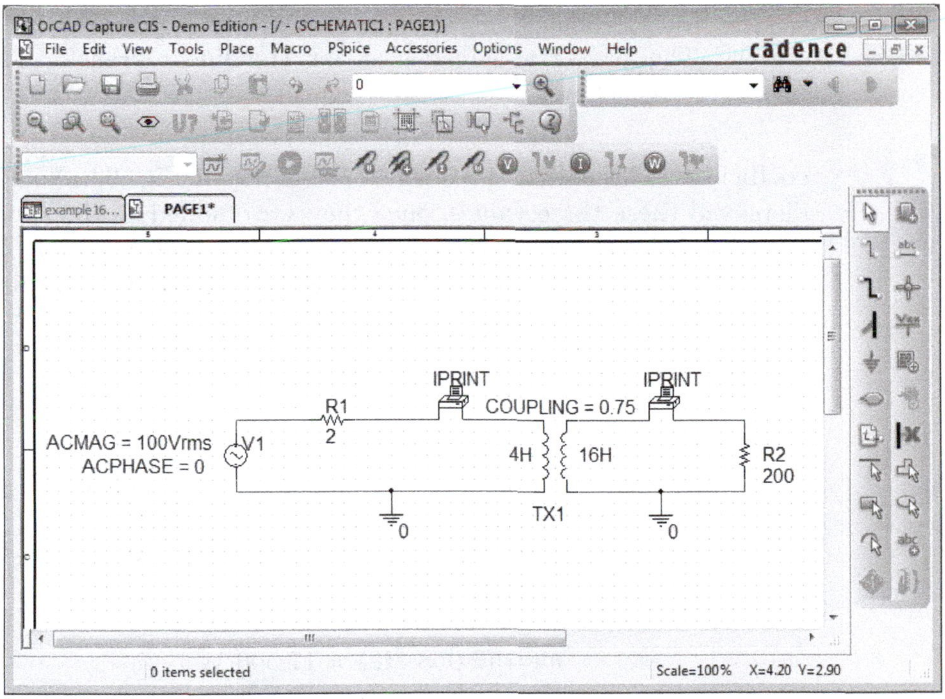

Figure 92: The schematic for the circuit in Fig. 91.

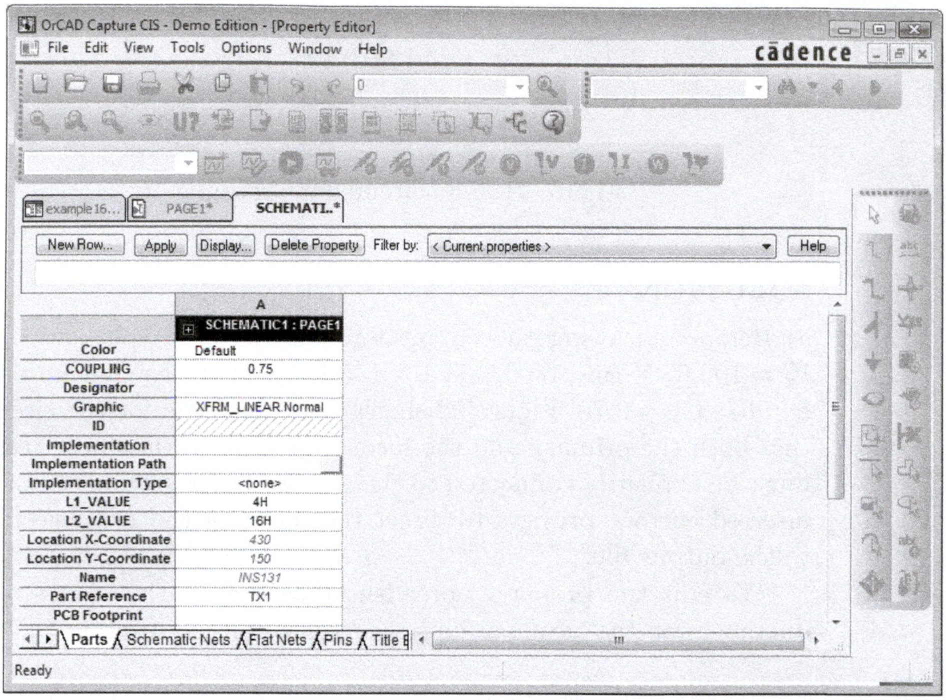

Figure 93: The property spreadsheet for the linear transformer in the schematic of Fig. 92.

coefficient. The edited spreadsheet is shown in Fig. 93. Note that we displayed these three values, once they were added to the spreadsheet. We specified AC Sweep analysis at the single frequency 1.59155 Hz.
b) The PSpice output file is shown in Fig. 94. From this output file,

$$\begin{aligned} \mathbf{I}_1 &= 2.958\underline{/-67.43^\circ}\ \text{A(rms)}; \\ \mathbf{I}_2 &= 0.6929\underline{/-16.09^\circ}\ \text{A(rms)}. \end{aligned}$$

Using the PSpice solution, we find that the total average power generated is

$$P_{\rm g} = (100)(2.958)\cos 67.43^\circ = 113.532\ \text{W}.$$

The average power dissipated in R_1 is

$$P_1 = (2.958)^2(2) = 17.500\ \text{W}.$$

The average power dissipated in R_2 is

$$P_2 = (0.6929)^2(200) = 96.022\ \text{W}.$$

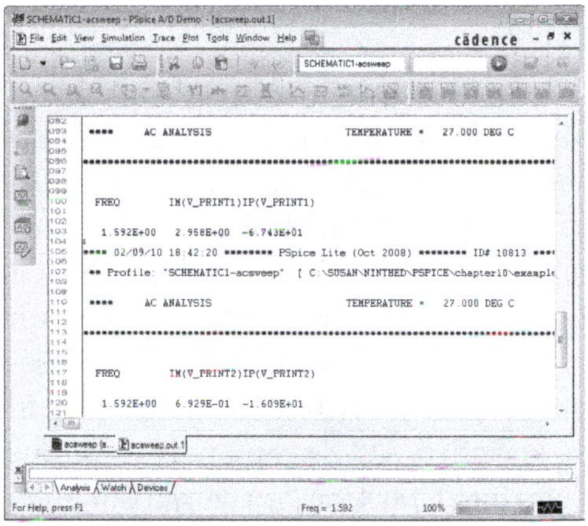

Figure 94: The output from AC Sweep analysis of the schematic in Fig. 91.

As $P_g = P_1 + P_2$, the PSpice solution is consistent with the conservation-of-energy principle.

10.2 IDEAL TRANSFORMERS

You can use the **XFRM_Linear** part to model the ideal transformers in a PSpice source file by making L_1 and L_2 large enough that ωL_1 and ωL_2 are much greater than the other impedances in the circuit. In addition to setting large values for L_1 and L_2, set the coefficient of coupling equal to 1. You bring the turns ratio of the ideal transformer into the model by using the relationship $L_1/L_2 = (N_1/N_2)^2$. This is illustrated in Example 17.

EXAMPLE 17

Find the magnitude and phase angle of the primary and secondary voltages for the circuit in Fig. 95, where $v_g = 50\cos 1000t$ V.

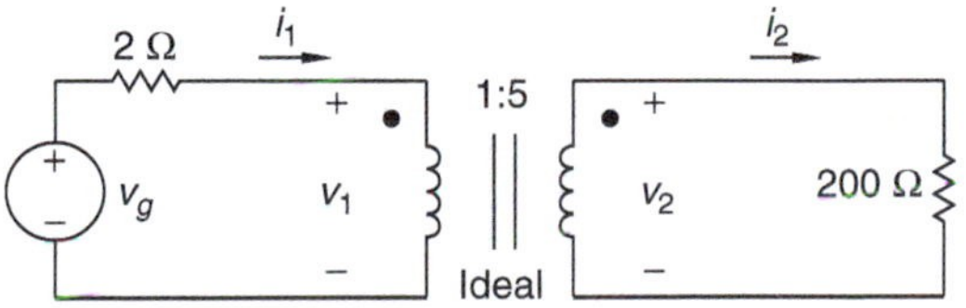

Figure 95: The circuit for Example 17.

SOLUTION

From the circuit diagram, $N_1/N_2 = 1/5$ and, therefore, $L_2 = 25L_1$. In picking values for L_1 and L_2, we want $\omega L_2 \gg 200\ \Omega$. As $\omega = 1000$ rad/s, if we make $L_2 = 200$ H, ωL_2 will be three orders of magnitude greater than R_2. Therefore, we pick $L_2 = 200$ H, and thus $L_1 = 8$ H.

We represented the circuit in Fig. 95 as a schematic in Fig. 96. Note that both sides of the transformer are connected to ground, and that we have inserted printers which will cause the voltages at the nodes of attachment to be written to the output file after PSpice analysis. As before, we specified AC Sweep analysis for a single frequency of 159.1549 Hz. The pertinent portion of the output file is shown in Fig. 97. Hence

$$\mathbf{V}_1 = 40\underline{/0.01146^\circ}\text{ V} \quad \text{and} \quad \mathbf{V}_2 = 200\underline{/0.01146^\circ}\text{ V}.$$

You should verify that the analytic solution for $\mathbf{V}_1$ and $\mathbf{V}_2$ yields

$$\mathbf{V}_1 = 40\underline{/0^\circ}\text{ V} \quad \text{and} \quad \mathbf{V}_2 = 200\underline{/0^\circ}\text{ V}.$$

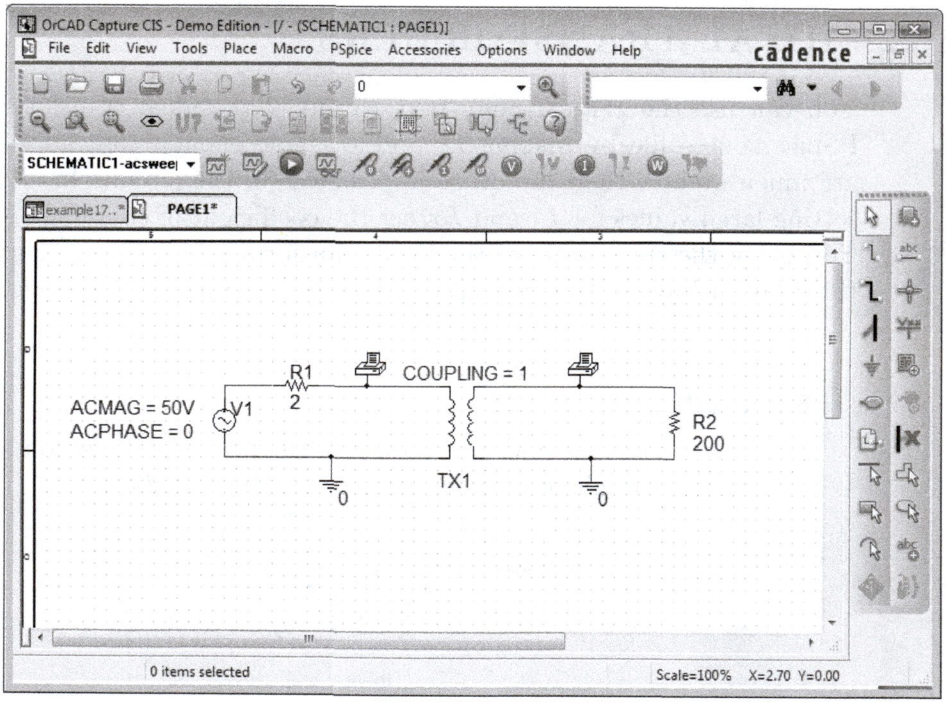

Figure 96: The schematic for the circuit in Fig. 98.

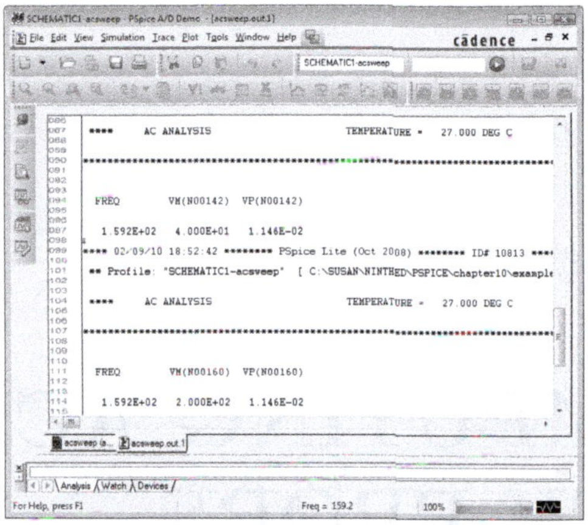

Figure 97: The output from AC Sweep analysis of the schematic in Fig. 96.

ASSESSMENT

- Use PSpice to solve the following problems: 9.77(a), 10.35(a), 10.36(c), and 10.41(a).

Chapter 11

COMPUTING AC POWER WITH PROBE

The interactive graphics capabilities provided by plotting routines offer the opportunity to study the relationships among the various ac power quantities. In this chapter, we investigate the relationships among current, voltage, instantaneous power, average power, and reactive power for the circuit shown in Fig. 98. Although we introduce several new features of plotting routines, you do not need any new PSpice constructs in order to investigate ac power.

To begin, construct a schematic for the parallel RC circuit in Fig. 98, as shown in Fig. 99. Specify transient analysis of this circuit for 100 μs.

When the blank graph appears with the main menu below it, select the menu item Plot/Add Plot to Window three times to create four blank plots on the screen. You can plot many different variables in these four plots and compare them. First, select the top plot on the screen, and plot the voltage at v_1. The plot should look like the first one (top) shown in Fig. 100. Next, select the second plot on the screen, and plot the current through the source, i_s. Note the phase relationship between the source voltage and current: The current leads the voltage because the load is capacitive. Compute the power

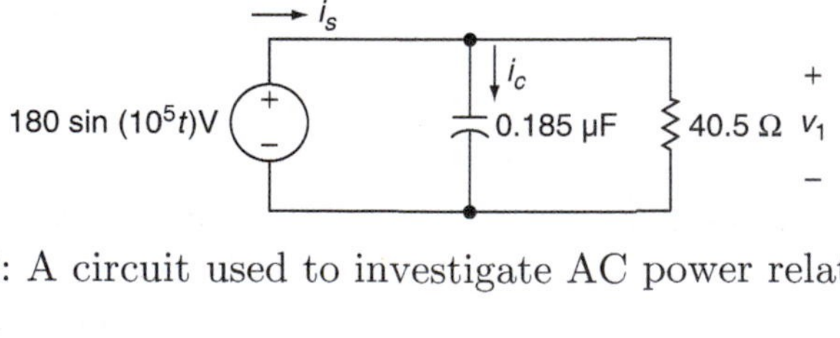

Figure 98: A circuit used to investigate AC power relationships.

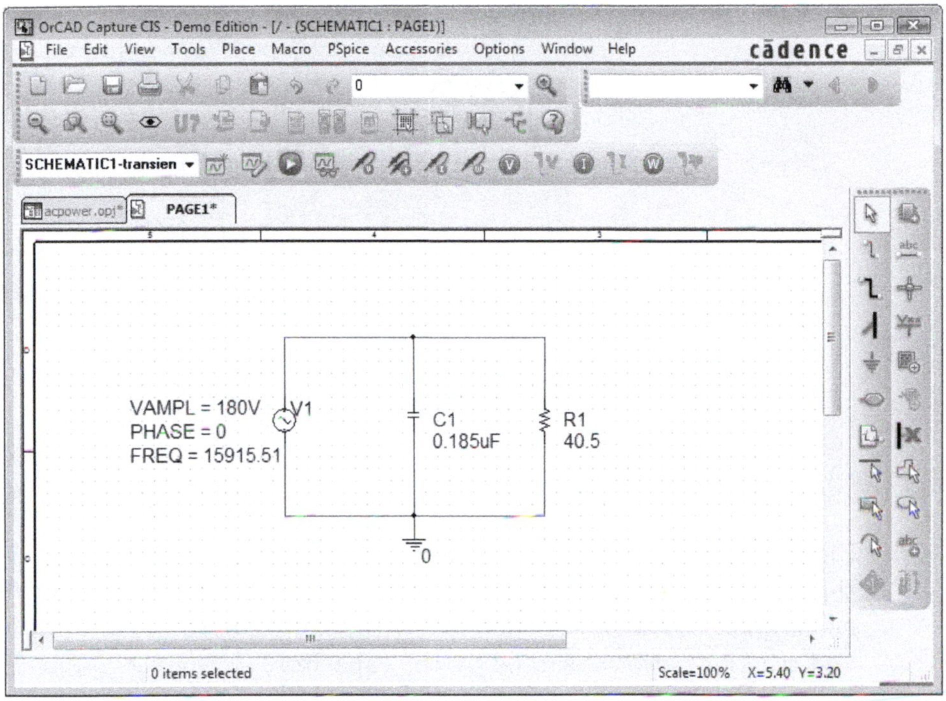

Figure 99: The schematic for the circuit in Fig. 98.

factor angle of the load as follows:

$$
\begin{aligned}
Z_{\text{L}} &= R \| 1/j\omega C \\
&= \frac{R}{j\omega RC + 1}; \\
\underline{/Z_{\text{L}}} &= 0 - \arctan(\omega RC) \\
&= -\arctan(10^5 \cdot 40.5 \cdot 0.185 \times 10^{-6}) \\
&= -36.84^\circ.
\end{aligned}
$$

Now, select the third plot, and graph the instantaneous power supplied by the source. Recall that the expression for instantaneous power is $v(t) \cdot i(t)$. Probe already has a variable that represents the instantaneous power of every component, so just select `-W(V1)`. Note that the minus sign (−) inverts the sign of the instantaneous power to conform to the PSpice sign convention. In this plot, you have confirmed that the frequency of the instantaneous power is double the frequency of the voltage and current.

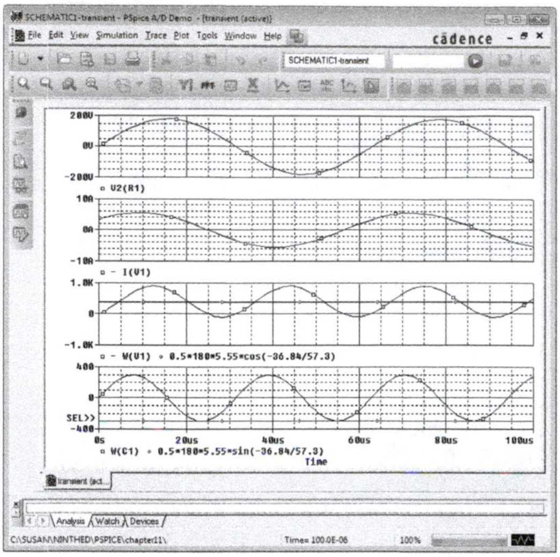

Figure 100: Some ac power characteristics from the circuit shown in Fig. 98: top plot, source voltage; second plot, source current; third plot, instantaneous and average power supplied by the source; bottom plot, instantaneous and reactive power absorbed by the capacitor.

Now display the average power supplied by the source on top of the plot of the instantaneous power. The formula for average power is

$$P = \frac{1}{2} V_{\text{m}} I_{\text{m}} \cos(\underline{/Z_{\text{L}}}).$$

Select the top two plots, one at a time, and use the cursor to find the maximum values $V_{\text{m}} = 180$ and $I_{\text{m}} = 5.55$. Select the third plot again, and add the trace described by the expression

```
.5*180*5.55*cos(-36.84/57.3)
```

Note that we divided the impedance angle ($\underline{/Z_{\text{L}}} = -36.84°$) by 57.3 to convert the angle from degrees to radians, as required by the plotting routines. In the third plot shown in Fig. 100, the equation for average power produced a straight line that runs through the center of the plot of instantaneous power, as you might have expected. You can use the cursor to find the value of this average power is 400 W.

Use the fourth plot to examine the reactive power in the circuit. On the fourth plot, generate a plot of the instantaneous power absorbed by the capacitor. Note that this instantaneous power has an average value of zero. On top of this plot, place the plot of the reactive power supplied by the source, which is described by the expression.

```
.5*180*5.55*sin(-36.84/57.3)
```

You have completed the plot shown in Fig. 100. Now try to describe the relationship between the instantaneous and average reactive power absorbed by the capacitor.

This exploration of the ac power characteristics of the circuit shown in Fig. 98 using the power of the plotting routines necessarily has been brief. You should take the time to investigate further; try to construct plots of complex power, for example.

ASSESSMENT

- Use PSpice to solve the following problems: 10.5, 10.18(a), and 10.62(a).

Chapter 12

FREQUENCY RESPONSE

We turn our attention to the constructs in PSpice that permit you to examine how a circuit behaves as a function of the frequency of its input. This behavior is the frequency response of the circuit.

12.1 SPECIFYING FREQUENCY VARIATION AND NUMBER

You use AC Sweep analysis, introduced in Section 9.2 for sinusoidal steady-state analysis, to obtain the frequency response of a circuit. You have a choice of frequency variations: linear or logarithmic. If you choose logarithmic, you must then specify decade or octave. Recall that a decade is a 10-to-1 change in frequency and that an octave is a 2-to-1 change in frequency. When you select the type of frequency variation, you must also select the number of frequencies (points) if the variation is linear or the number of frequencies (points) per decade (octave) if the frequency variation is decade (octave). As before, you must also specify a start and end frequency for the sweep, using the units Hz.

If you select a linear variation, the frequency increment Δf is

$$\Delta f = \frac{\text{FSTOP} - \text{FSTART}}{\text{NP} - 1}.$$

For example, if FSTOP = 1801 Hz, FSTART = 1 Hz, and the number of points is 181, then

$$\Delta f = \frac{1801 - 1}{181 - 1} = 10 \text{ Hz}.$$

If you select a decade variation, the frequencies within a given decade are

$$f_k = \mathrm{FSD} \cdot 10^{k/\mathrm{ND}}$$

where FSD is the frequency at the start of the decade, ND is the number of points per decade, and k is an integer in the range from 1 to ND. For example, if ND = 20, FSTART = 50, and FSTOP = 5000, the frequencies in the decade between 50 and 500 would be

$$f_k = (50)10^{k/20}$$

where $k = 1, 2, 3, \ldots, 20$. Hence

$$\begin{aligned} f_1 &= (50)10^{1/20} = 56.10 \text{ Hz}; \\ f_2 &= (50)10^{2/20} = 62.95 \text{ Hz}; \\ &\vdots \end{aligned}$$

If you select an octave variation, the frequencies within a given octave are

$$f_k = \mathrm{FSO} \cdot 2^{k/\mathrm{NO}}$$

where FSO is the frequency at the start of the octave, NO is the number of points per octave, and k is an integer in the range from 1 to NO. For example, if FSTART = 400 Hz, FSTOP = 3200 Hz, and NO = 10, the frequencies in the octave from 400 to 800 Hz would be

$$f_k = (400)2^{k/10}.$$

Hence

$$\begin{aligned} f_1 &= (400)2^{0.1} = 428.71 \text{ Hz}; \\ f_2 &= (400)2^{0.2} = 459.48 \text{ Hz}; \\ &\vdots \end{aligned}$$

12.2 FREQUENCY RESPONSE OUTPUT

Most often you will want to view the results of an AC Sweep analysis using plots. But you can also use printer parts to cause the results of the analysis to be written to the output file. You may specify the voltage and current variables in rectangular, polar, or decibel values by placing a Y in the appropriate column on the property spreadsheet for the printer.

Example 18 illustrates how to analyze the frequency response of a parallel RLC circuit with PSpice.

EXAMPLE 18

a) The current source in the circuit shown in Fig. 101 is $50 \cos \omega t$ mA. Use Probe to plot v_o versus f from 1000 to 2000 Hz in increments of 10 Hz on a linear frequency scale.

b) From the Probe plot, estimate the resonant frequency, the bandwidth, and the quality factor of the circuit.

c) Compare the results obtained in (b) with an analytic solution for f_o, β, and Q.

SOLUTION

a) The PSpice schematic for the circuit in Fig. 101 is shown in Fig. 102. We specify the parameters of the AC Sweep as shown in Fig. 103.

b) Figure 104 shows the Probe plot of v_o versus f. We need to choose a linear range for the x-axis such as the one shown in Fig. 104. Using the Probe Cursor, we note that the peak amplitude of 400 V occurs at a frequency of 1590.2 Hz in Fig. 104(a). Thus we estimate the resonant frequency at 1590 Hz.

To estimate the bandwidth, we use both cursors to find the frequencies where $v_o = 400/\sqrt{2} = 282.84$ V. From the Probe plot in Fig. 104(b), the closest values are 282.731 V at 1552.1 Hz and 282.731 V at 1632.0 Hz. The bandwidth is then shown as the variable "dif" in the Probe Cursor Window, and has the value 79.839, or about 80 Hz.

We calculate the quality factor from the relationship $Q = f_o/\beta = 1590/80 = 19.88$.

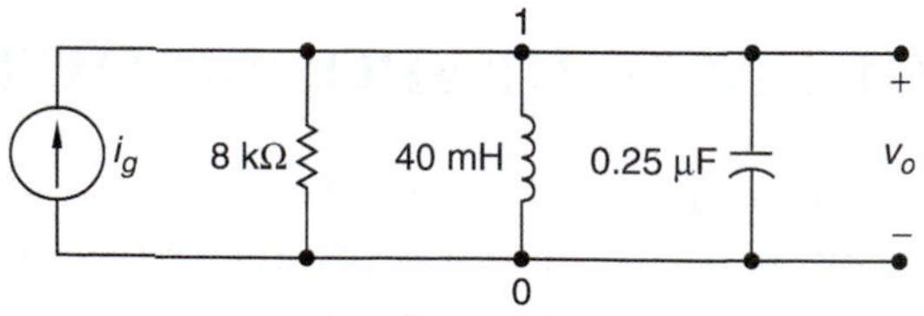

Figure 101: The circuit for Example 18.

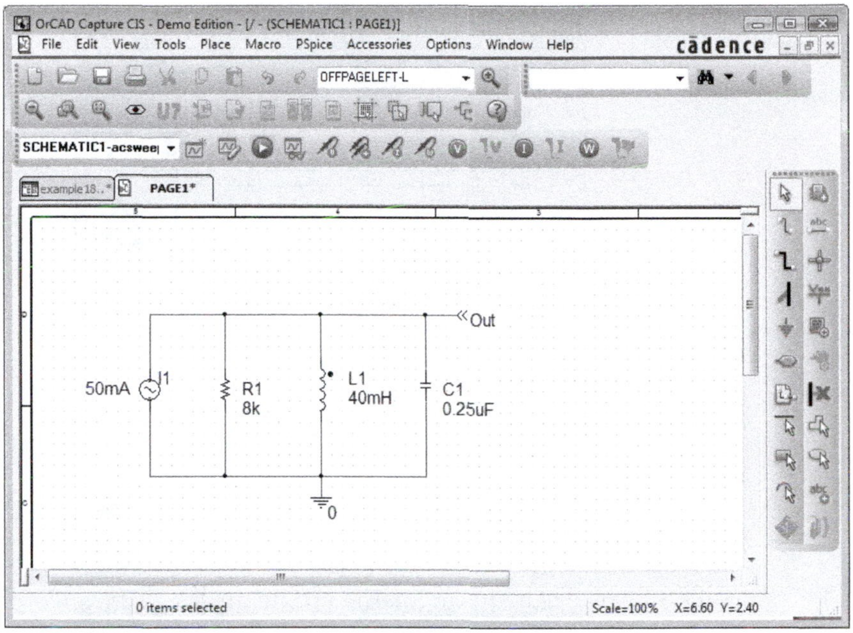

Figure 102: The schematic for the circuit in Fig. 101.

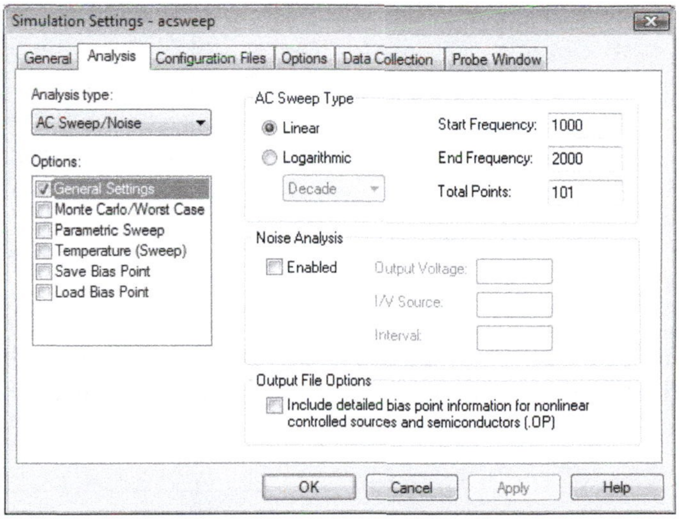

Figure 103: Specifying the parameters for the AC Sweep of the schematic in Fig. 102.

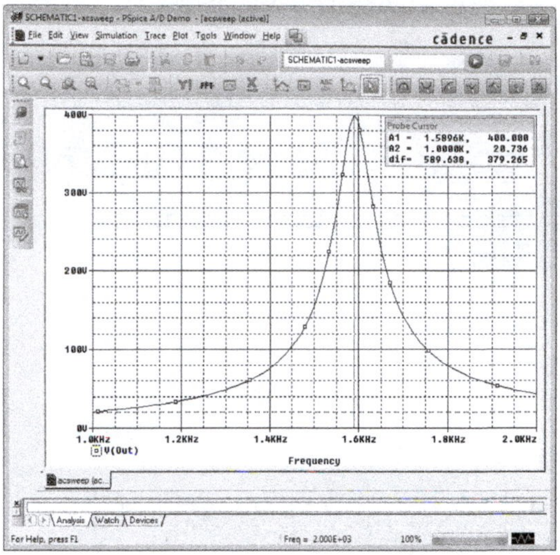

(a)

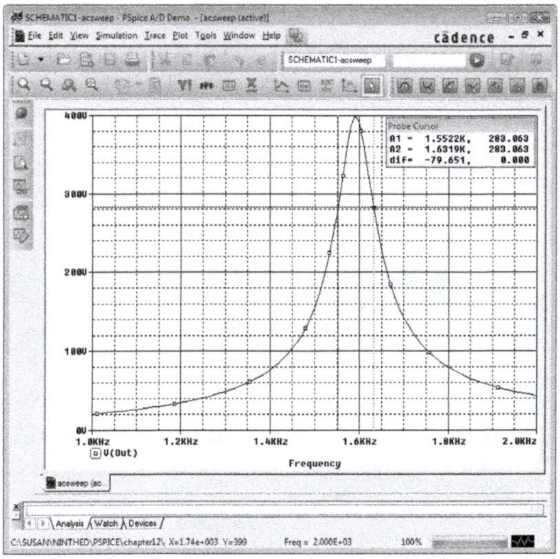

(b)

Figure 104: Probe plot of v_o versus f for Example 18, using the cursor in (a) to identify the resonant frequency and in (b) to identify the bandwidth.

c) A direct analysis of the circuit yields

$$\begin{aligned} f_\mathrm{o} &= 1591.55 \text{ Hz;} \\ f_1 &= 1552.26 \text{ Hz;} \\ f_2 &= 1631.84 \text{ Hz;} \\ \beta &= f_2 - f_1 = 79.58 \text{ Hz;} \\ Q &= f_\mathrm{o}/\beta = 20. \end{aligned}$$

We summarize the comparison with the PSpice analysis as follows:

Quantity	Analysis	PSpice
f_o	1591.55	1589.6
f_1	1552.26	1552.2
f_2	1631.84	1631.9
β	79.58	79.651
Q	20.00	19.96

If necessary, we can obtain a more accurate analysis around the resonant frequency from PSpice by rerunning the program with different values of the Start and Stop frequencies and a larger number of points.

In Example 19, we use a global parameter to vary the value of the capacitor in the circuit shown in Fig. 101. We use Probe to look at the effect of this variation on the frequency response characteristics of the circuit.

EXAMPLE 19

Modify the PSpice schematic for Example 18 to step the capacitor values from 0.15 μF through 0.35 μF in increments of 0.5 μF. Then use Probe to display the frequency response characteristics for all values of capacitance. Comment on the effect of the changing capacitance.

SOLUTION

The modified schematic is shown in Fig. 105. Note that we have replaced the single-valued capacitor with a capacitor whose value is {CVAL}. We have also added a Param part with a new attribute called CVAL.

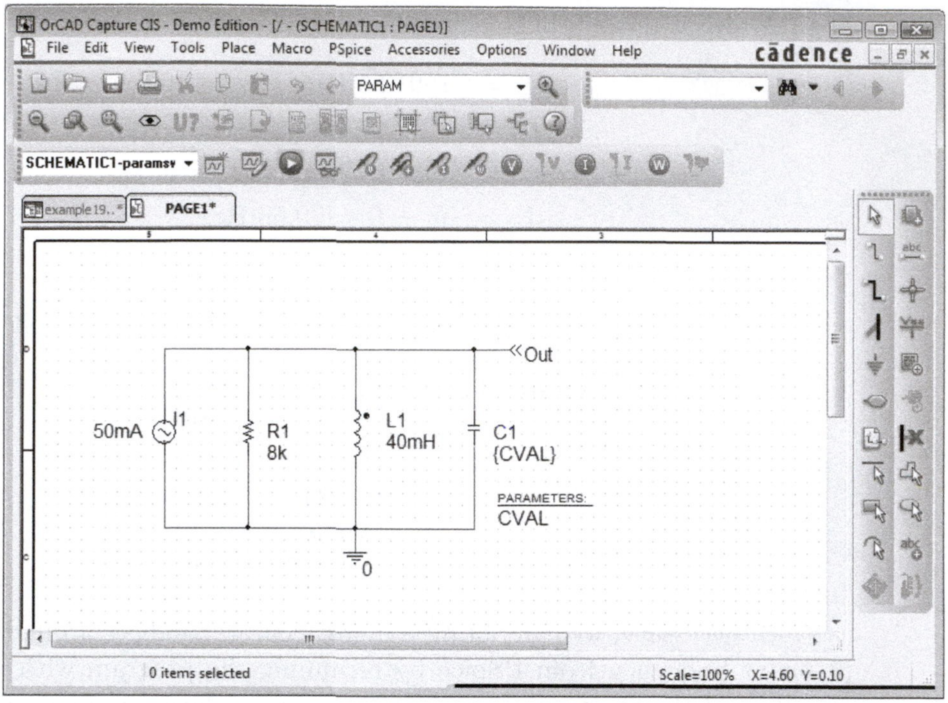

Figure 105: The schematic for the circuit in Fig. 101.

Then we have edited the analysis specification in several ways. First, we changed the AC Sweep to start at 500 Hz and end at 2500 Hz, to accommodate the broader frequency band that results from the range of capacitor values. Second, we have added a Parametric Sweep, to vary the global parameter CVAL from 0.15 μF to 0.35 μF in steps of 0.05 μF.

Figure 106 shows the Probe plot. The smallest value of capacitance produced the plot farthest to the right. As the capacitance increases, the plots move to the left. Hence the resonant frequency *decreases* as the capacitance *increases*. But we expect this result because the equation for resonant frequency for an *RLC* circuit is

$$\omega_r = \sqrt{\frac{1}{\mathrm{LC}}}.$$

Furthermore, as the capacitance increases, the resonant peak becomes sharper, as you can see in Fig. 106. That is, as the capacitance increases, the quality becomes higher. This result, too, comes as no

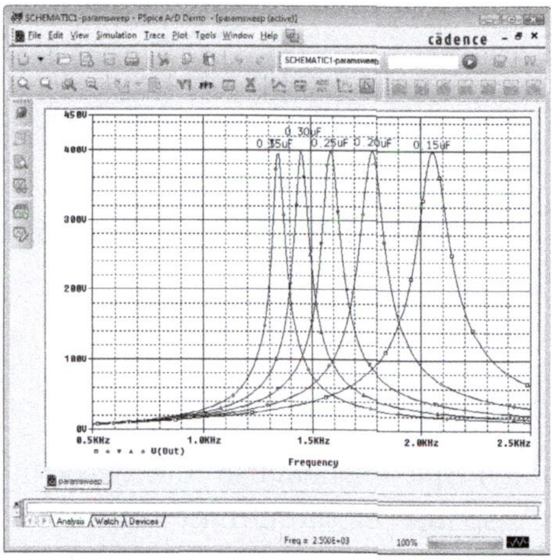

Figure 106: The relationship between changing capacitance and circuit frequency response.

surprise because the equation for Q in a parallel RLC circuit is

$$Q = \mathrm{R}\sqrt{\frac{\mathrm{C}}{\mathrm{L}}}.$$

12.3 BODE PLOTS WITH PROBE

In Section 12.2, we described the method for generating a frequency response plot with PSpice and Probe. The voltage and current PRINT parts for frequency response data can provide you with output voltages and currents in several different forms: real and imaginary parts, magnitude and phase angles, and dB magnitude and phase angles. Probe also allows you to plot frequency response data by using these same forms. In Section 12.2, we plotted linear voltage magnitude versus linear frequency. We now turn to another form: dB voltage magnitude versus log frequency and phase angle versus log frequency. This is the form used to generate Bode plots.

Example 20 explores the frequency response of a series RLC circuit for varying values of resistance. This example presents the frequency response in Bode plot form.

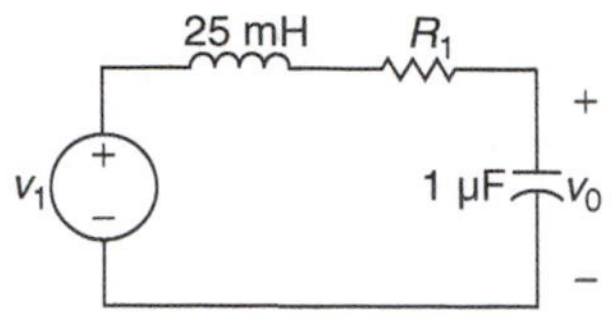

Figure 107: The circuit for Example 20.

EXAMPLE 20

Construct a PSpice schematic and associated analysis to generate the frequency response of the circuit shown in Fig. 107 for three different values of resistance: 5 Ω, 50 Ω, and 500 Ω. Then use Probe to plot the output voltage magnitude in dB and output voltage phase angle versus log frequency.

SOLUTION

In order to analyze the circuit three times with three different values of resistance, we use a resistor with value {RVAL}. We also add a global parameter part to our schematic and add an RVAL attribute, which becomes the value of the resistor. To generate Bode plots, make sure that the magnitude of the source voltage is 1, and its phase angle is 0°. The resulting schematic is shown in Fig. 108.

We need to perform some preliminary analysis of the circuit to generate the values used in specifying the AC Sweep analysis: the starting and ending frequencies. We choose them so that the frequency band of interest, that is, the frequency band that contains the natural frequency of the circuit, is centered between the two end-point frequencies.

Because this is a series *RLC* circuit, we know that the center frequency and the bandwidth are given by

$$\omega_o = \frac{1}{\sqrt{\mathrm{LC}}} \quad \text{and} \quad \beta = \frac{\mathrm{R}}{\mathrm{L}}.$$

For the circuit shown in Fig. 107, the center frequency is

$$\begin{aligned} \omega_o &= \frac{1}{\sqrt{(25 \times 10^{-3})(1 \times 10^{-6})}} = 6324.56 \text{ rad/sec} \\ &\approx 1000 \text{ Hz}, \end{aligned}$$

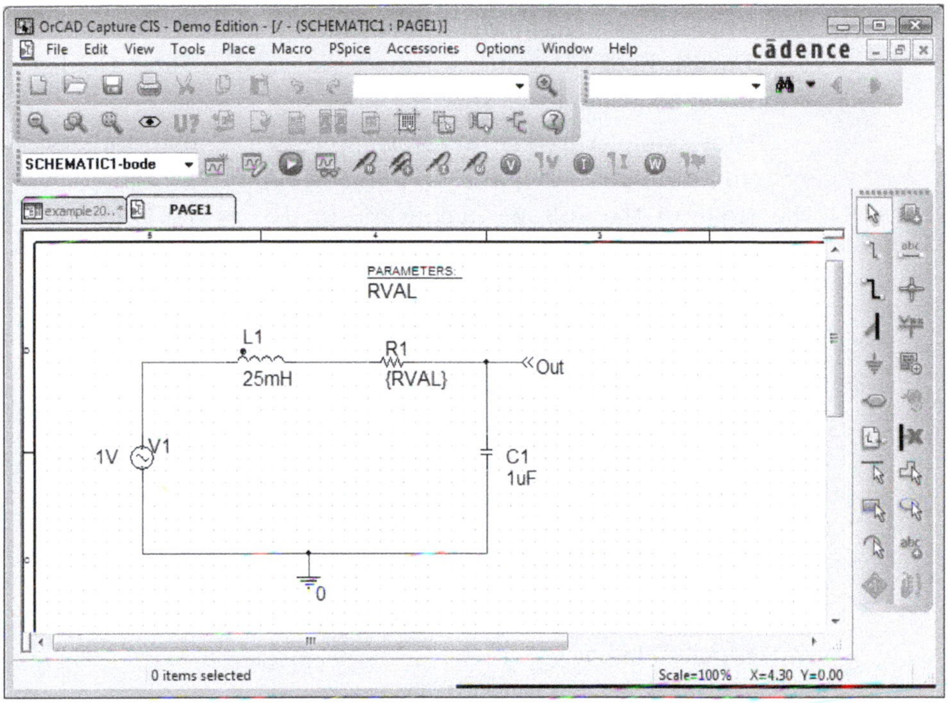

Figure 108: Schematic for the circuit in Fig. 107.

and the bandwidth ranges from

$$\begin{aligned} \beta_{\text{narrow}} &= \frac{5}{25 \times 10^{-3}} = 200 \text{ rad/sec} \\ &\approx 32 \text{ Hz} \end{aligned}$$

to

$$\begin{aligned} \beta_{\text{wide}} &= \frac{500}{25 \times 10^{-3}} = 20000 \text{ rad/sec} \\ &\approx 3.2 \text{ kHz.} \end{aligned}$$

Thus we choose 100 Hz as the starting frequency and 10,000 Hz as the ending frequency, which places the center frequency in the middle on a logarithmic frequency scale and provides adequate room for the wideband circuit's response.

After entering these values in the AC Sweep dialog, check the Parameter Sweep box so that we can perform this ac steady-state analysis for each of the three resistor values. This time, instead of specifying a range of resistor values, we create a list of values. To do this, click on the List button in the Parameter Sweep dialog, and enter the list of values, separated by commas or spaces.

After completing the analysis, we enter Probe and use the menu system to create two plots on the screen. In the upper plot, we add the trace of the voltage magnitude in dB at the Out node, which is the magnitude of the voltage across the capacitor in dB. Note that the frequency axis is already logarithmic because the AC Sweep analysis specified frequencies in decades. In the lower plot, we add the trace of the voltage phase angle at the Out node, which is the phase angle of the voltage across the capacitor.

The transfer function for this circuit takes the form

$$H(s) = \frac{1/\mathrm{LC}}{s^2 + (\mathrm{R}/\mathrm{L})s + 1/\mathrm{LC}}$$

and so has two poles and no zeros. As the value of R changes, the values of the poles change

R	First pole	Second pole
5	$-100 + j6323.76$	$-100 - j6323.76$
50	$-1000 + j6245.00$	$-1000 - j6245.00$
500	-2254.03	-17745.97

The two smaller values of R have complex poles with damping coefficients of 0.0158 for R = 5 ohms and 0.1581 for R = 50 ohms.

Figure 109 presents the Bode plot for the three cases. The upper plot is the magnitude plot in dB, which for each case starts at 0 dB for low frequencies and has a slope of −40 dB/decade for high frequencies. The top dB curve is for 5 ohms and the bottom curve is for 500 ohms. As expected, the dB curve peaks upward near 1000 Hz for cases with complex poles and low damping coefficients.

The lower plot in Figure 109 is the phase plot. Each curve ranges between 0° at low frequencies and −180° at high frequencies. All curves go through −90° at the center frequency near 1000 Hz. The steeper curves correspond to complex poles and lower damping coefficients.

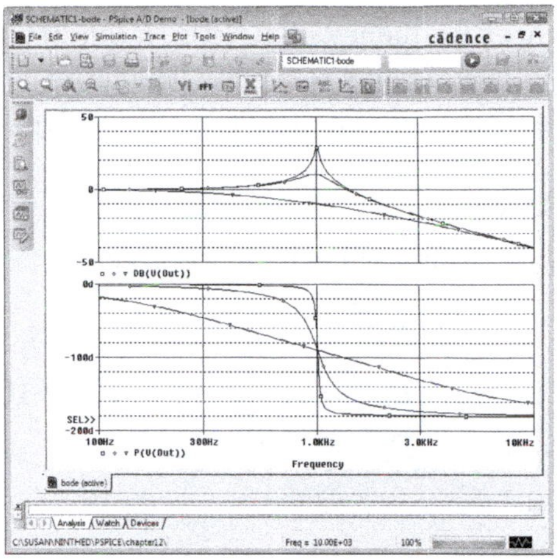

Figure 109: The frequency response for the circuit shown in Fig. 107.

12.4 FILTER DESIGN

The examples in this chapter thus far have illustrated the use of PSpice in analyzing the frequency response of circuits. We can also employ PSpice and Probe to assist us in designing op amp-based active filters, whose specifications are usually centered on meeting certain frequency domain requirements. Example 21 demonstrates the use of PSpice and Probe in verifying the behavior of a high-Q bandpass filter.

EXAMPLE 21

The circuit in Fig. 110 is an active high-Q bandpass filter. Using 1 nF capacitors and an ideal op amp, design values for the three resistors to

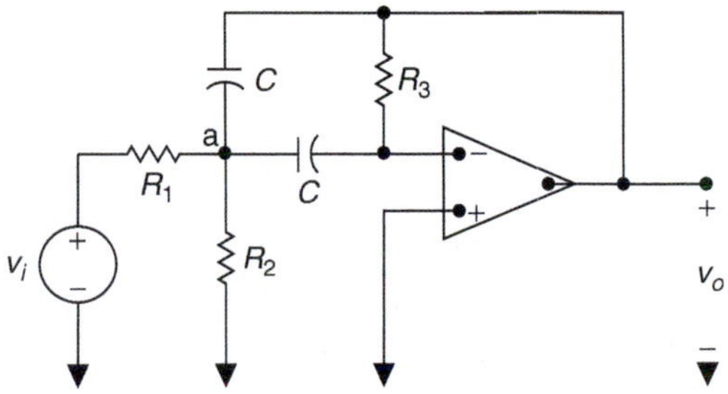

Figure 110: The active high-Q bandpass filter for Example 21.

yield a center frequency of 10 kHz, a quality factor of 10, and a passband gain of 3. Use Probe to verify that the resistor values you compute produce a filter that satisfies the three frequency response specifications.

SOLUTION

The resistor design equations are given by

$$\begin{aligned} R_1 &= \mathrm{Q/K} = 10/3 = 3.333; \\ R_2 &= \mathrm{Q}/(2\mathrm{Q}^2 - \mathrm{K}) = 10/197; \\ R_3 &= 2\mathrm{Q} = 20. \end{aligned}$$

The scaling factors are $k_\mathrm{f} = 2\pi(\omega_\mathrm{o}) = 20000\pi$ and $k_\mathrm{m} = 1/(Ck_\mathrm{f}) = 10^9/k_\mathrm{f}$. After scaling,

$$\begin{aligned} R_1 &= 53.1\ \mathrm{k\Omega} \\ R_2 &= 808\ \Omega \\ R_3 &= 318.3\ \mathrm{k\Omega} \end{aligned}$$

The PSpice schematic for the design is given in Fig 111. Note the use of the PSpice model for the ideal op amp. The Probe plot of the frequency response of the simulated circuit is shown in Fig. 112. We have used the cursors to verify the bandwidth of 10000/10 = 1000 Hz, and you can further see from this plot that the center frequency and passband gain specifications are also satisfied. You might want to explore the robustness of this design by substituting a realistic op amp model for the ideal model used and checking to see if the design specifications are still met.

ASSESSMENT

- Use PSpice to find and plot the frequency response for the voltage or current indicated in the following problems, and answer the questions posed: 14.22, 14.32(a)--(d), 15.28(a),(f), and 15.29(a),(f).

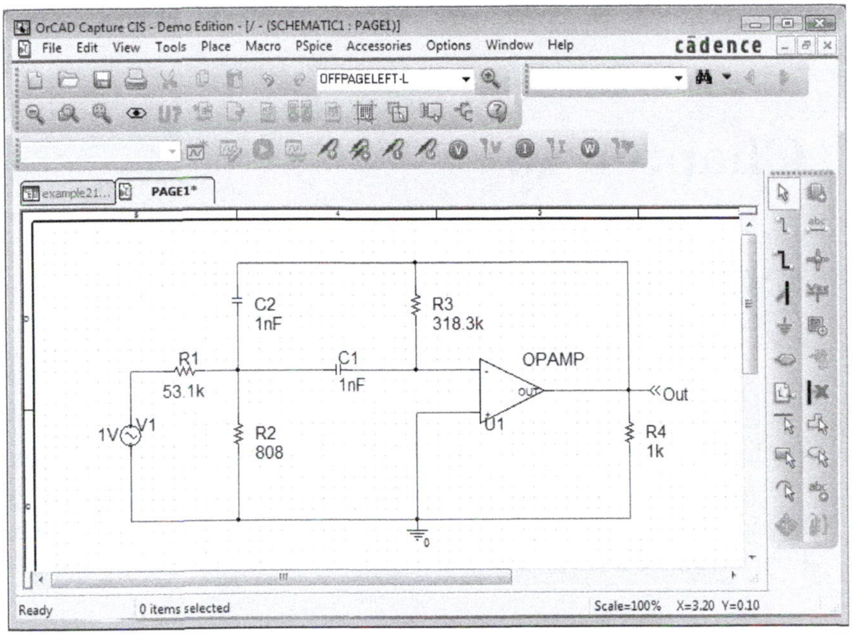

Figure 111: The PSpice schematic for the high-Q filter design.

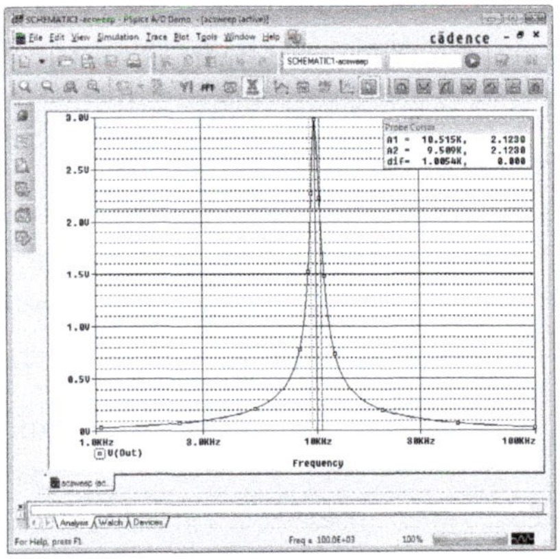

Figure 112: The frequency response of the high-Q filter design for Example 21.

Chapter 13

FOURIER SERIES

Both PSpice and Probe can help you understand the frequency spectrum characteristics of periodic waveforms. We begin with a brief discussion of PSpice models for periodic waveforms, then turn to PSpice support for Fourier analysis.

13.1 PULSED SOURCES

The PSpice model for a periodic pulse voltage source has the part name Vpulse, in the Source library. The pulse parameters are

$$\begin{aligned}
\text{V1} &= \text{initial value (volts);} \\
\text{V2} &= \text{pulsed value (volts);} \\
\text{TD} &= \text{delay time (seconds);} \\
\text{TR} &= \text{rise time (seconds);} \\
\text{TF} &= \text{fall time (seconds);} \\
\text{PW} &= \text{pulse width (seconds);} \\
\text{PER} &= \text{period (seconds).}
\end{aligned}$$

These parameters are set in the property spreadsheet for the pulsed source. Figure 113 shows a graph of the periodic pulse.

We illustrate the analysis of a circuit with a pulsed source in the next section.

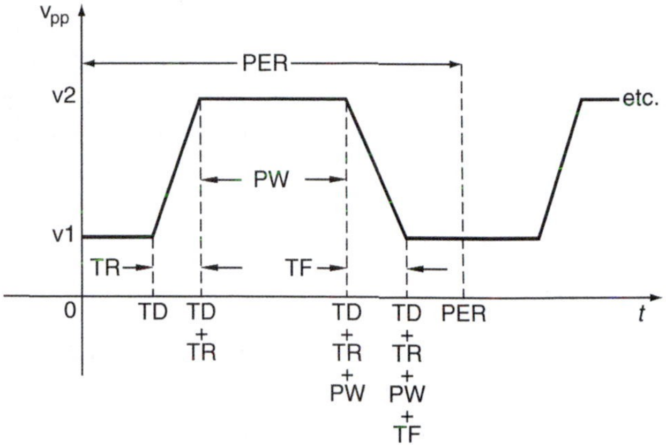

Figure 113: A periodic pulse voltage source.

13.2 FOURIER ANALYSIS

Transient analysis must precede Fourier series analysis because the Fourier series coefficients are computed for the last complete cycle for the transient analysis waveform. Therefore, the duration of the transient analysis must be at least one cycle of the waveform. If you want PSpice to compute the Fourier series coefficients of the steady-state waveform, the transient analysis should be performed for a number of cycles sufficient to ensure the decay of the natural response.

Fourier analysis is specified as an option in transient analysis, as we will see in Example 22. You will be asked to supply the value of the fundamental frequency of the periodic signal in hertz; recall that the fundamental frequency in hertz is the inverse of the period of the signal in seconds. You will also be asked to identify the variables for which the Fourier coefficients are computed. As usual, these variables must be voltages or currents. You will also be asked how many harmonic terms should be written to the output file. Including a printer part is not necessary since the Fourier coefficients automatically print in the output file as PSpice analysis proceeds.

If a PSpice has performed transient analysis, you may use Probe to calculate and display the Fourier transform coefficients, whether or not you have asked for the Fourier coefficients to be written to the output file. In fact, Probe calculates the Fourier transform coefficients independent of any PSpice calculation of Fourier series coefficients. The frequency resolution displayed by Probe is inversely proportional to the extent in time of the transient analysis. Therefore, if you want a more accurate Fourier transform plot, you should specify a longer time for the transient analysis. You reach the Fourier transform mode in Probe by choosing the appropriate submenu.

Example 22 illustrates a pulsed voltage source, the PSpice computation of the Fourier series coefficients, and the Probe computation of the Fourier transform coefficients.

EXAMPLE 22

Suppose that the pulsed waveform shown in Fig. 114(a) is input to the *RC* circuit shown in Fig. 114(b). Use a Fourier series analysis of the input waveform to predict the frequency content of the output waveform for two values of C: 1 μF and 10 μF. Then use the Fourier transform mode in Probe to confirm the PSpice analysis and to display the frequency spectra graphically.

SOLUTION

We begin by computing the Fourier series components for the pulsed voltage signal shown in Fig. 114(a). Note that this waveform has two properties of symmetry: odd symmetry and half-wave symmetry. These symmetry properties tell us several things about the Fourier coefficients:

- The dc component is 0.
- The Fourier series has only sine terms; the a_n coefficients are 0 for all values of n.

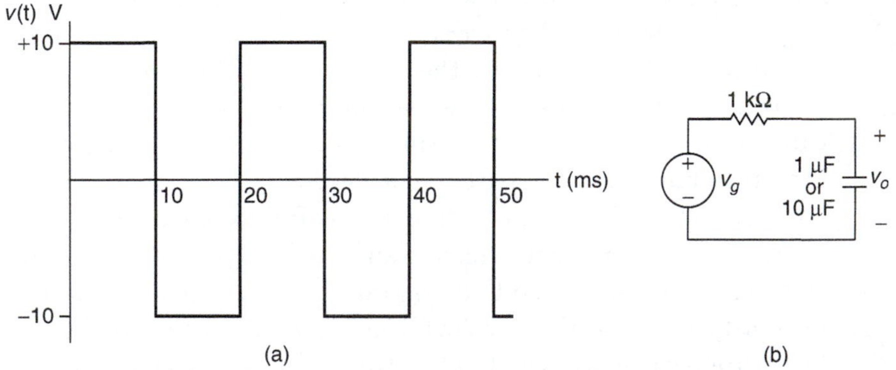

Figure 114: An *RC* circuit with pulsed voltage input; (a) the pulsed voltage waveform; (b) the circuit.

- Only the odd sine terms are nonzero; the b_n coefficients are 0 for even n.
- Because of the form of the a_n's and the b_n's, the Fourier coefficients expressed in magnitude and phase-angle form are

  $$A_n = b_n \quad \text{and} \quad \theta_n = -90^\circ, \quad \text{for odd } n.$$

  Both the magnitude and phase angles are zero for the even harmonics.

Because of waveform symmetry, we can simplify the formula for the coefficients:

$$b_n = \frac{4}{T}\int_0^{T/2} f(t)\sin k\omega_o t dt = \frac{80}{2k\pi} = \frac{40}{k\pi}, \quad k \text{ odd}.$$

Thus we can represent the input to the circuit in Fig. 114(b) as an infinite series of the form

$$v_g = \frac{40}{\pi}\sin\omega_o t + \frac{40}{3\pi}\sin 3\omega_o t + \frac{40}{5\pi}\sin 5\omega_o t + \ldots$$

The circuit shown in Fig. 114(b) is linear and thus superposition applies. To compute the response of this circuit to the pulsed waveform, we consider the response to each term in the Fourier series independently and sum those responses. From a phasor analysis of the circuit, we know that

$$V_{ok} = \frac{V_{gk}}{1 + jk\omega_o \text{RC}},$$

which we represent in magnitude and phase-angle form as

$$V_{ok} = \frac{\frac{40}{k\pi}\angle 0^\circ}{\sqrt{(k\omega_o\text{RC})^2 + 1^2}\angle \arctan k\omega_o\text{RC}},$$

The following table compares the magnitude and phase angle of the input pulsed waveform with the magnitude and phase angle of the voltage across the capacitor for the circuit with C = 1 μF and the circuit with C = 10 μF, for the first three odd harmonic frequencies. For the circuit with the smaller capacitor, the magnitudes do not change much between the input and the output. For the circuit with the larger capacitor, the magnitudes are much smaller at the output than they are at the input at all harmonic frequencies. We use Probe to investigate this difference further.

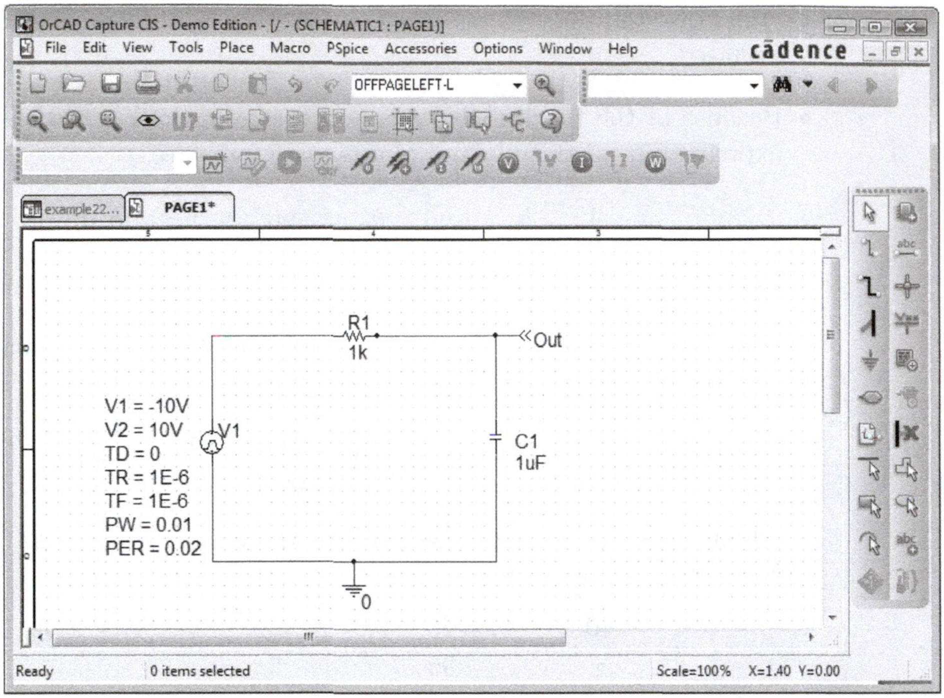

Figure 115: The schematic for the circuit shown in Fig. 114(b), with C = 1 μF.

Harmonic	V_g	V_o, **C** $= 1\ \mu$**F**	V_o, **C** $= 10\ \mu$**F**
First	$12.7\angle 0^\circ$	$12.1\angle -17.4^\circ$	$3.9\angle -72.3^\circ$
Third	$4.2\angle 0^\circ$	$3.1\angle -43.3^\circ$	$0.4\angle -83.9^\circ$
Fifth	$2.5\angle 0^\circ$	$1.4\angle -57.5^\circ$	$0.2\angle -86.4^\circ$

The PSpice schematic that describes the circuit shown in Fig. 114(b) with the smaller capacitor value and the input shown in Fig. 114(a) is shown in Fig. 115. The spreadsheet containing the values of the parameters specifying the pulsed voltage source is shown in Fig. 116. You should confirm that the values of the properties match the pulsed waveform in Fig. 114(a).

Next, we specify the parameters for transient analysis. The choice of a value for TSTOP is based on an analysis of the time constant for this *RC* circuit, which for C = 1 μF is 1 ms and for C = 10 μF is 10 ms. Thus, setting TSTOP to 60 ms will allow a 6 time constant decay of the natural response for the larger value of C, ensuring that the Fourier series coefficients will be computed for the steady-state waveform.

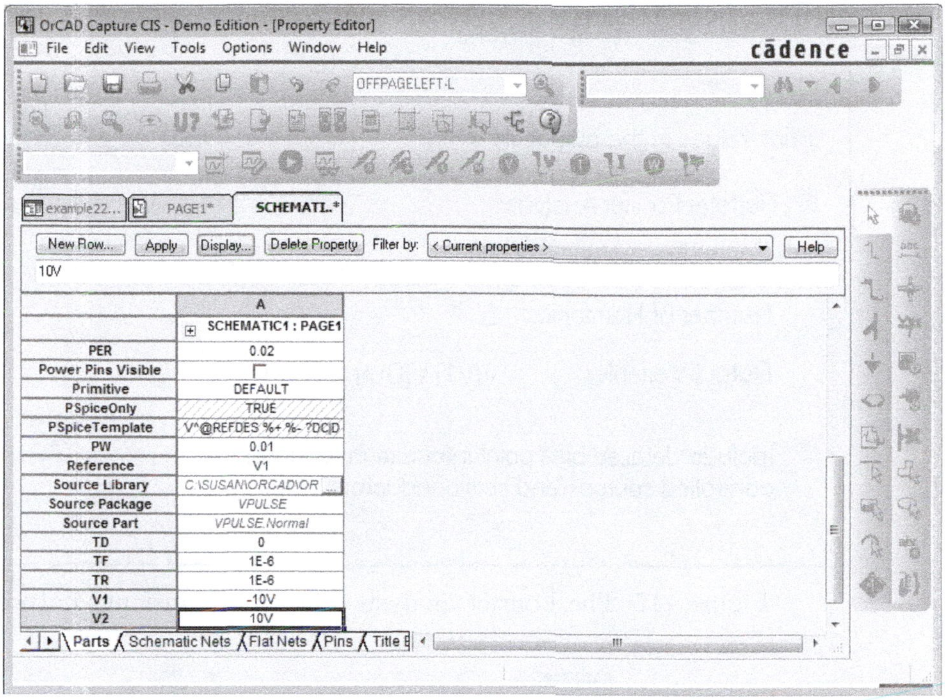

Figure 116: The spreadsheet used to specify the properties of the pulsed voltage source.

We direct PSpice to compute the Fourier coefficients and write them to the output file by clicking the Output File Options button in the Transient Analysis dialog. Now we check the box to perform the Fourier analysis and specify the fundamental frequency, the number of harmonics, and the output variables. The completed dialog for Fourier analysis is shown in Fig. 117.

The Fourier coefficients for the first five harmonics for the output capacitor voltage are shown in Fig. 118. Note that this PSpice analysis confirms our previous hand calculations for the first three odd harmonics. Note further that the even harmonics are essentially zero valued; the nonzero values that are printed arise because of the imprecision inherent in digital computer calculations.

Now examine the graph of the time response for both the input pulsed waveform and the output capacitor voltage produced by Probe shown in Fig. 119. Because of the small capacitor value, the output voltage basically is able to follow the input voltage. Finding that the Fourier coefficients of the input and output voltages are so similar is not surprising.

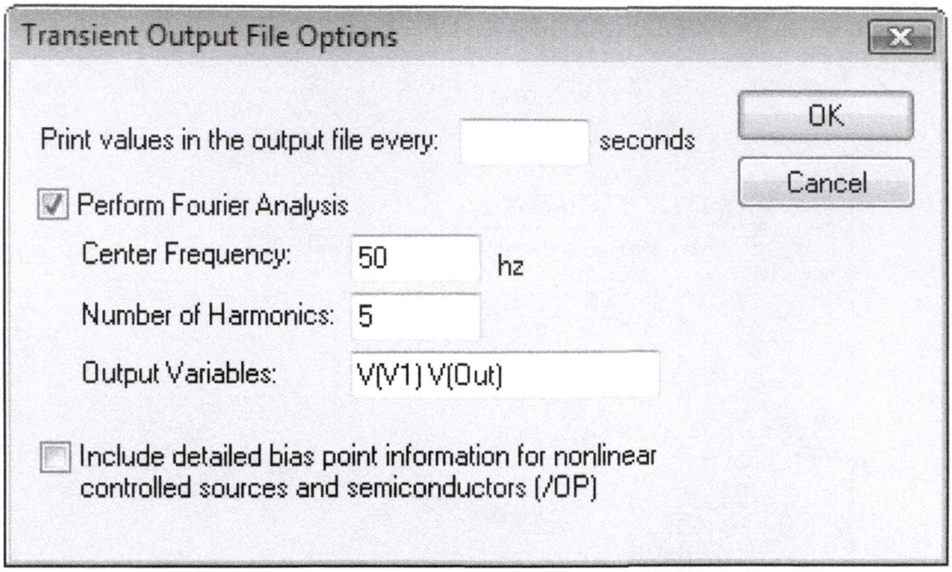

Figure 117: The Fourier analysis dialog in Transient analysis.

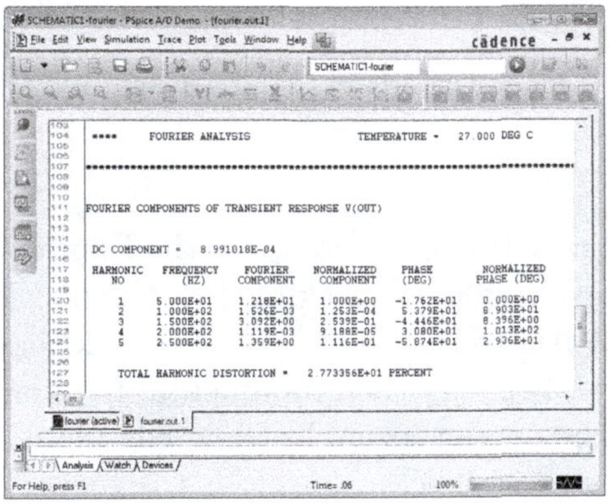

Figure 118: The first five harmonics for the output waveform for the circuit simulated by the schematic in Fig. 115.

Figure 120, produced from Fig. 119 in Probe by clicking the Toolbar button labeled FFT and changing the range of the x-axis, graphically confirms this similarity.

We changed the capacitor value in the schematic (Fig. 115) to 10 μF, and repeated the analysis. The resulting Fourier series coefficients for the first five harmonics of the capacitor voltage, confirming the hand calculations presented, are shown in Fig. 121.

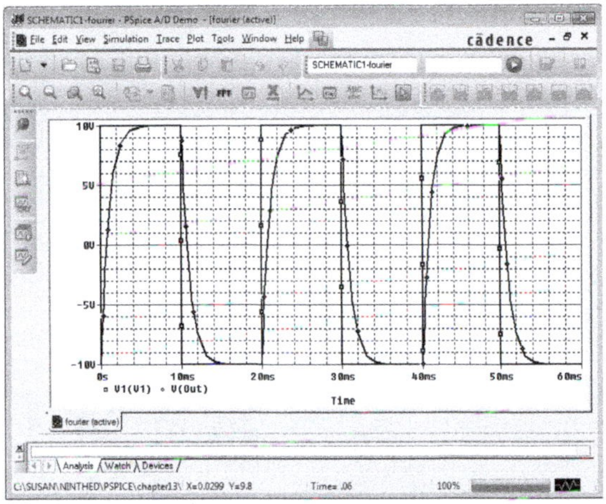

Figure 119: Probe time response plot for the *RC* circuit with small C.

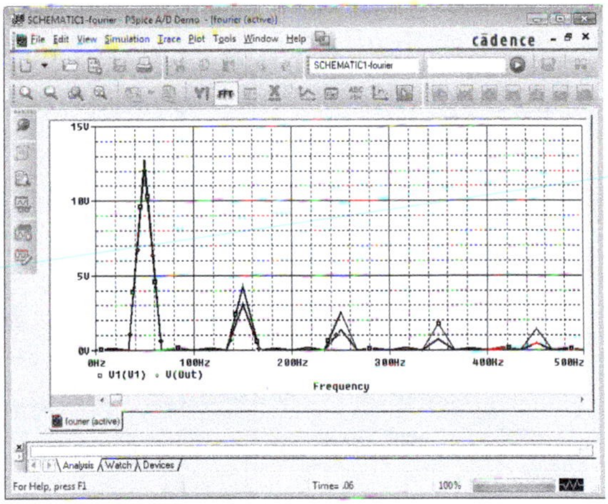

Figure 120: Probe Fourier transform of the waveforms shown in Fig. 119.

Probe plots of the time response for the input and output voltages, as shown in Fig. 122, illustrate that the large capacitor value prevents the output voltage from following the input voltage. Because the voltage waveforms are so different in the time domain, you should expect very different Fourier coefficients. You saw this difference in both the hand calculations and the PSpice analysis. The Probe-generated plot of the Fourier transformed voltages shown in Fig. 123 further confirms the different spectral content in the two waveforms.

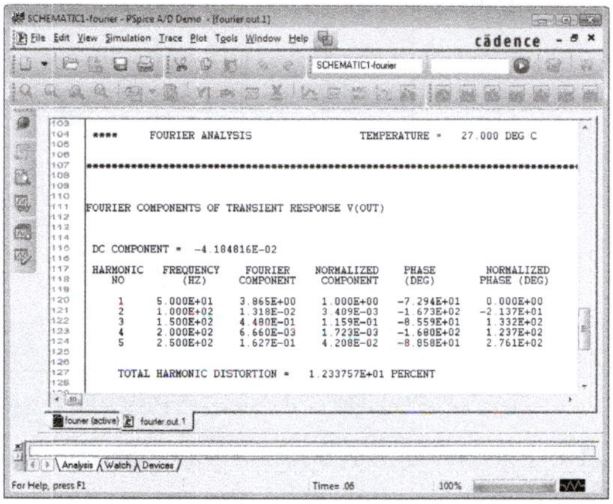

Figure 121: The first five harmonics for the output waveform for the circuit in Fig. 114 with C = 10 μF.

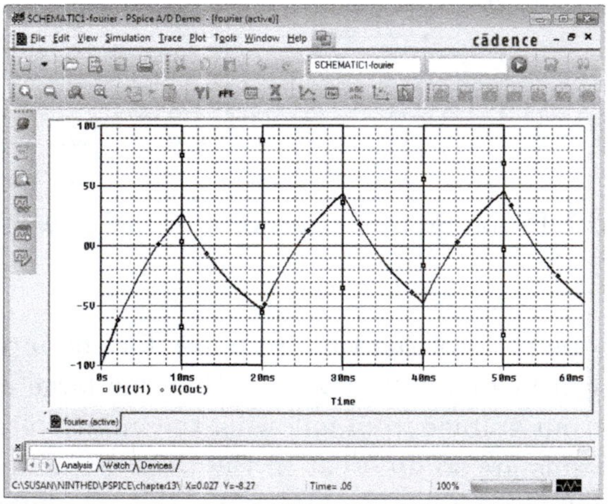

Figure 122: Probe time response plot for the *RC* circuit with C = 10 μF.

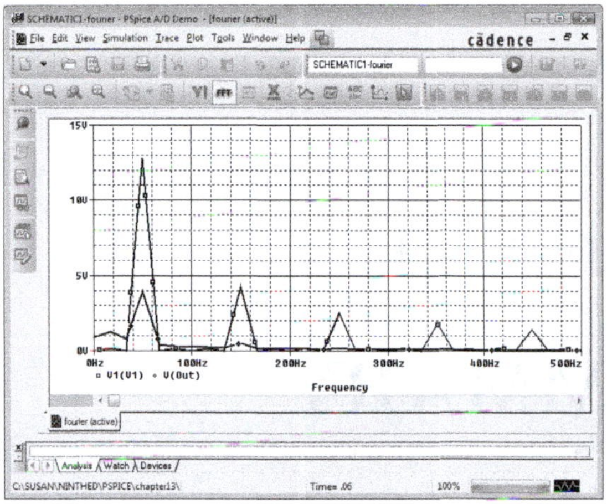

Figure 123: Probe Fourier transform of the waveforms shown in Fig. 122.

ASSESSMENT

- Use PSpice to find and plot the Fourier coefficients for the voltages and currents indicated in the following problems: 16.27 and 16.28.

Chapter 14

SUMMARY

OrCAD PSpice A/D is one of numerous software packages designed to simulate an electric circuit or network. One of the helpful features of these simulation programs is that they allow the user to define the circuit in terms of a topological, as opposed to a mathematical, description. Although the primary purpose of PSpice is to facilitate the analysis of circuits containing nonlinear electronic components, we have introduced it to you as a tool for analyzing linear, lumped-parameter circuits. There are several reasons for this early introduction. First, it acquaints you with a powerful computational tool. Second, you can test the computer-generated values by direct computation, thus gaining confidence in your ability to formulate correct source files. Third, the computer makes it possible for you to attack circuit problems that are computationally very complex and lengthy if attempted without the aid of a computer. These complex problems often add significantly to your understanding of circuit behavior. Finally, simulation software is part of modern engineering practice. The introduction of such software at an elementary level is a first step in preparing you for what lies ahead.

TO EXPLORE FURTHER

Additional information is available by using the Help menus, accessing the README file on the enclosed disk, and following the links from this file to many useful websites.

APPENDIX

QUICK REFERENCE TO PSPICE NETLIST STATEMENTS

Before PSpice can begin analysis, the schematic must be translated into a collection of statements which identify the types of components, their attributes, and their interconnections. This collection is the netlist. This appendix presents the netlist syntax for most of the schematic parts presented in this supplement. Being familiar with the netlist syntax is helpful when debugging errors in your schematic, and is necessary when creating new subcircuit models or modifying existing subcircuit models for parts.

DATA STATEMENTS

In the syntax for data statements, *xxx* represents any alphanumeric string used to identify uniquely a circuit component, n+ represents the positive circuit node, and n− represents the negative circuit node. Positive voltage drop is from n+ to n−, and positive current flows from n+ to n−. Any elements in the syntax specification that are optional are enclosed by brackets ([]).

INDEPENDENT DC VOLTAGE AND CURRENT SOURCES

SYNTAX (VOLTAGE) Vxxx n+ n− DC value

SYNTAX (CURRENT) Ixxx n+ n− DC value

where

value is the dc voltage in volts for the voltage source and the dc current in amperes for the current source.

DESCRIPTION Provides a constant source of voltage or current to the circuit.

EXAMPLES `V_in $N_0001 $N_0002 DC 3.5`

`I_source $N_0003 0 DC 0.25`

INDEPENDENT AC VOLTAGE AND CURRENT SOURCES

SYNTAX (VOLTAGE) Vxxx n+ n− AC mag [phase]

SYNTAX (CURRENT) Ixxx n+ n− AC mag [phase]

where

mag is the magnitude of the ac waveform in volts for the voltage source and in amperes for the current source;

phase is the phase angle of the ac waveform in degrees and has a default value of 0.

DESCRIPTION Supplies a sinusoidal voltage or current at a fixed frequency to the circuit.

EXAMPLES `V_ab $N_0003 $N_0002 AC 10 90`

`I_12 $N_0004 $N_0006 AC 0.5`

INDEPENDENT TRANSIENT VOLTAGE AND CURRENT SOURCES

SYNTAX (VOLTAGE) Vxxx n+ n− trans_type

SYNTAX (CURRENT) Ixxx n+ n− trans_type

where trans_type is one of the following transient waveform types:

- *Exponential* `EXP` (start peak [delay1] [rise] [delay2] [fall])

  where

  start is the initial value of voltage in volts or current in amperes;

  peak is the maximum value of voltage in volts or current in amperes;

  delay1 is the delay in seconds prior to a change in voltage or current from the initial value, which is 0 seconds by default;

rise is the time constant in seconds of the exponential decay from start to peak, which has a default value of tstep seconds (see `.TRAN`);

delay2 is the delay in seconds prior to introducing a second exponential decay, from the maximum value toward the initial value, which has a default value of delay1 + tstep seconds (see `.TRAN`);

fall is the time constant in seconds of the exponential decay from peak to final, which has a default value of tstep seconds (see `.TRAN`).

- *Pulsed* `PULSE`(min max [delay] [rise] [fall] [width] [period])

  where

  min is the minimum value of the waveform in volts for the voltage source and in amperes for the current source;

  max is the maximum value of the waveform in volts for the voltage source and in amperes for the current source;

  delay is the time in seconds prior to the onset of the pulse train, which has a default of 0 seconds;

  rise is the time in seconds for the waveform to transition from min to max, which has a default value of tstep seconds (see `.TRAN`);

  fall is the time in seconds for the waveform to transition from max to min, which has a default value of tstep seconds (see `.TRAN`);

  width is the time in seconds that the waveform remains at the max value, which has a default value of tstop seconds (see `.TRAN`);

  period is the time in seconds that separates the pulses in the pulse train, which has a default of tstop seconds (see `.TRAN`).

- *Piecewise linear* `PWL`(t1 val1 t2 val2 . . .)

  where

  t1 val1 are a paired time (in seconds) and value (in volts for a voltage source and amperes for a current source) that specifies a corner of the waveform; all pairs of times and values are linearly connected to form the whole waveform.

- *Damped sinusoid* `SIN`(off peak [freq] [delay] [damp] [phase]) where

off is the initial value of the voltage in volts or the current in amperes;

peak is the maximum amplitude of the voltage in volts or the current in amperes;

freq is the sinusoidal frequency in hertz, which has a default value of 1/tstop hertz (see `.TRAN`);

delay is the time in seconds that the waveform stays at the initial value before the sinusoidal oscillation begins, which has a default value of 0 seconds;

damp is the sinusoidal damping factor in seconds^{-1} used to specify the decaying exponential envelope for the sinusoid, which has a default value of 0 seconds^{-1};

phase is the initial phase angle of the sinusoidal waveform in degrees, which has a default value of 0 degrees.

DESCRIPTION Supplies a time-varying voltage or current to the circuit whose waveform can be characterized as exponential, pulsed, piecewise linear, or damped sinusoidal.

EXAMPLES

```
V_s  $N_0003 $N_0001 EXP(2 6 .5 .1 .5 .2)

I_in  $N_0004 0 PULSE(-.3 .3 0 .01 .01 1 2)

I_5  $N_0002 $N_0003 PWL(0 .2 1 .6 1.5 6 3 -.5)

V_g  0  $N_0003 SIN(0 2 100 0 0 90)
```

DEPENDENT VOLTAGE-CONTROLLED VOLTAGE SOURCE

SYNTAX Exxx   n+   n−   cn+   cn−   gain

where

cn+ is the positive node for the controlling voltage;

cn− is the negative node for the controlling voltage;

gain is the ratio of the source voltage (between n+ and n−) to the controlling voltage (between cn+ and cn−).

DESCRIPTION Provides a voltage source whose value depends on a voltage measured elsewhere in the circuit.

EXAMPLE `E_op $N_0001 $N_0002 $N_0004 0 .5`

DEPENDENT CURRENT-CONTROLLED CURRENT SOURCE

SYNTAX Fxxx n+ n− Vyyy gain

where

Vyyy is the name of the voltage source through which the controlling current flows;

gain is the ratio of the source current (flowing from n+ to n−) to the controlling current (flowing through Vyyy).

DESCRIPTION Provides a current source whose value depends on the magnitude of a current flowing through a voltage source elsewhere in the circuit.

EXAMPLE `F_dep $N_0003 $N_0002 V_control 10`

DEPENDENT VOLTAGE-CONTROLLED CURRENT SOURCE

SYNTAX Gxxx n+ n− cn+ cn− gain

where

cn+ is the positive node for the controlling voltage;

cn− is the negative node for the controlling voltage;

gain is the ratio of the source current (flowing from n+ to n−) to the controlling voltage (between cn+ and cn−), in ohms^{-1}.

DESCRIPTION Provides a current source whose value depends on the magnitude of a voltage measured elsewhere in the circuit.

EXAMPLE `G_on $N_0003 $N_0006 $N_0004 $N_0001 0.35`

DEPENDENT CURRENT-CONTROLLED VOLTAGE SOURCE

SYNTAX Hxxx n+ n− Vyyy gain

where

Vyyy is the name of the voltage source through which the controlling current flows;

gain is the ratio of the source voltage (between n+ and n−) to the controlling current (flowing through Vyyy), in ohms.

DESCRIPTION Provides a voltage source whose value depends on the magnitude of current flowing through another voltage source elsewhere in the circuit.

EXAMPLE `H_out $N_0007 $N_0002 V_dummy -2.5E-3`

RESISTOR

SYNTAX Rxxx n+ n− [mname] value

where

mname is the name of a `RES` model defined in a `.MODEL` statement (see `.MODEL`)---note that mname is optional;

value is the resistance, in ohms.

DESCRIPTION Models a resistor, a circuit element whose voltage and current are linearly dependent.

EXAMPLE `R_for $N_0003 $N_0004 16E3`

INDUCTOR

SYNTAX Lxxx n+ n− [mname] value [IC = icval]

where

mname is the name of an `IND` model defined in a `.MODEL` statement (see `.MODEL`)---note that mname is optional;

value is the inductance, in henries;

icval is the initial value of the current in the inductor, in amperes--note that specifying the initial condition is optional.

DESCRIPTION Models an inductor, a circuit element whose voltage is linearly dependent on the derivative of its current.

EXAMPLE `L_44 $N_0001 $N_0009 3E-3 IC = 1E-2`

CAPACITOR

SYNTAX Cxxx n+ n− [mname] value [IC = icval]

where

mname is the name of a `CAP` model defined in a `.MODEL` statement (see `.MODEL`)---note that mname is optional;

value is the capacitance, in farads;

icval is the initial value of the voltage across the capacitor, in volts---note that specifying the initial condition is optional.

DESCRIPTION Models a capacitor, a circuit element whose current is linearly dependent on the derivative of its voltage.

EXAMPLES
```
C_two $N_0004 0 cmod 2E-6
.MODEL cmod CAP(C = 1)
```

MUTUAL INDUCTANCE

SYNTAX Kxxx Lyyy Lzzz value

where

Lyyy is the name of the inductor on the primary side of the coil;

Lzzz is the name of the inductor on the secondary side of the coil;

value is the mutual coupling coefficient, k, which has a value such that $0 \leq k < 1$.

DESCRIPTION Models the magnetic coupling between any two inductor coils in a circuit.

EXAMPLE `K_ab L_a L_b 0.5`

SUBCIRCUIT DEFINITION

SYNTAX `.SUBCKT` name [nodes] `.ENDS`

where

name is the name of the subcircuit, as referenced by an `X` statement (see subcircuit call);

nodes is the optional list of nodes used to identify the connections to the subcircuit;

`.ENDS` signifies the end of the subcircuit definition.

DESCRIPTION Used to provide "subroutine"-type definitions of portions of a circuit. When the subcircuit is referenced in an `X` statement, the definition between the `.SUBCKT` statement and the `.ENDS` statement replaces the `X` statement in the source file.

EXAMPLES

```
.SUBCKT opamp $N_0001 $N_0002 $N_0003 $N_0004 $N_0005

.ENDS
```

SUBCIRCUIT CALL

SYNTAX Xxxx nodes name

where

nodes is the optional list of circuit nodes used to connect the subcircuit into the rest of the circuit; there must be as many nodes in this list as there are in the subcircuit definition (see "Subcircuit Definition");

name is the name of the subcircuit as defined in a `.SUBCKT` statement.

DESCRIPTION Replaces the `X` statement with the definition of a subcircuit, which permits a subcircuit to be defined once and used many times within a given source file.

EXAMPLE

```
X_amp $N_0004 $N_0002 $N_0008 $N_0005 $N_0001 opamp
```

LIBRARY FILE

SYNTAX `.LIB` [fname]

where

fname is the name of the library file containing `.MODEL` or `.SUBCKT` statements referenced in the source file, which by default is the nominal or evaluation library file.

DESCRIPTION Used to reference models or subcircuits.

EXAMPLE `.LIB mylib.lib`

DEVICE MODELS

SYNTAX `.MODEL` mname mtype [(par=value)]

where

mname is a unique model name, which is also used in the device statement that incorporates this model;

mtype is one of the model types available;

par=value is an optionally specified list of parameters and their assigned values, specific to the model type:

- `RES`, which models a resistor and has the parameter `R`, the resistance multiplier, whose default value is 1;
- `IND`, which models an inductor and has the parameter `L`, the inductance multiplier, whose default value is 1;
- `CAP`, which models a capacitor and has the parameter `C`, the capacitance multiplier, whose default value is 1.

DESCRIPTION Defines standard devices that can be used in a circuit and sets parameter values that characterize the specific device being modeled.

EXAMPLES `.MODEL lmodel IND(L = 2)`

```
.MODEL resist RES
```

CONTROL STATEMENTS

DC ANALYSIS

SYNTAX `.DC` [type] vname start end incre [nest-sweep]

where

type is the sweep type and must be `LIN` for a linear sweep from the start value to the end value, `OCT` for a logarithmic sweep from start value to end value in octaves, `DEC` for a logarithmic sweep from start value to end value in decades, or `LIST` if a list of values is to be used; the default sweep type is `LIN`;

vname is the name of the circuit element whose value is swept, usually an independent voltage or current source;

start is the beginning value of vname;

end is the final value of vname; note that when the sweep type is `LIST`, the start, end, and incre values are replaced by a list of values that vname will take on;

incre is the step size for the `LIN` sweep type and is the number of points per octave or decade for the `OCT` and `DEC` sweep types;

nest-sweep is an optional additional sweep of a second circuit variable, which follows the same syntax as the sweep for the first variable.

DESCRIPTION Provides a method for varying one or two dc source values and analyzing the circuit for each sweep value.

EXAMPLES `.DC V_ab $N_0001 $N_0010 .5`

```
.DC I_1 DEC .1 10 30 V_1 DEC 10 100 30
```

SENSITIVITY

SYNTAX `.SENS` vname

where vname is the circuit variable name (or list of circuit variable names) for which sensitivity analysis will be performed.

DESCRIPTION Computes and prints to the output file a dc sensitivity analysis of vname to the values of other circuit elements, including resistors, independent voltage and current sources, and voltage- and current-controlled switches.

EXAMPLE `.SENS V($N_0002) I(V_in)`

TRANSIENT ANALYSIS

SYNTAX `.TRAN` tstep tstop [npval] istep [`UIC`]

where

tstep is the time interval separating values that are printed or plotted;

tstop is the ending time for the transient analysis;

npval is the time between 0 and the first value printed or plotted, which by default is 0;

istep is the internal time step used for computing values, which by default is tstop/50;

`UIC` will bypass the calculation of the bias point, which usually precedes the transient analysis, and use the initial conditions specified by `IC =` in inductor and capacitor data statements instead.

DESCRIPTION Performs a transient analysis of the circuit described in the source file to calculate the values of circuit variables as a function of time.

EXAMPLE `.TRAN .01 10 0 .001 UIC`

INITIAL CONDITION

SYNTAX `.IC` Vnode=value

where

Vnode=value is one or a list of pairs consisting of a voltage node Vnode, represented in standard form, and the initial value in volts at that node.

DESCRIPTION Sets the initial conditions for transient analysis by specifying one or more node voltage values.

EXAMPLE `.IC V($N_0001) = 10`

AC ANALYSIS

SYNTAX `.AC` [type] num start end

where

type is the type of sweep, which must be one of the following keywords:

`LIN` for a linear sweep in frequency, which is the default;

`OCT` for a logarithmic sweep in frequency by octaves;

`DEC` for a logarithmic sweep in frequency by decades;

num is the total number of points in the sweep for a linear sweep and the number of points per octave or decade for a logarithmic sweep;

start is the starting frequency, in hertz;

end is the final frequency, in hertz.

DESCRIPTION Computes the frequency response of the circuit described in the source file as the frequency is swept either linearly or logarithmically from an initial value to a final value.

EXAMPLE `.AC LIN 300 10 1000`

STEPPED VALUES

SYNTAX `.STEP` [type] name start end incre

where

type is one of the following keywords describing the type of parameter variation:

`LIN` for a linear variation, which is the default;

`OCT` for a logarithmic variation by octaves;

`DEC` for a logarithmic variation by decades;

name is the name of the circuit element whose value is to be varied;

start is the starting value for the circuit element, in units appropriate to the type of circuit element;

end is the final value for the circuit element, in units appropriate to the type of circuit element;

incre is the step size for a linear variation and the number of values per octave or decade for the logarithmic variation. Note that when a discrete list of parameter values is desired, type is not used, and the name is followed by the keyword `LIST` and a list of parameter values.

DESCRIPTION Used to step a circuit element's value through a range either linearly or logarithmically or through a discrete list, analyzing the circuit for each value.

EXAMPLES `.STEP DEC RES RMOD(R) 10 10E4 3`

`.STEP V_2 LIST 1 4 12 19`

FOURIER SERIES ANALYSIS

SYNTAX `.FOUR` freq vname

where

freq is the fundamental frequency of the circuit;

vname is the variable name or the list of variable names for which Fourier series coefficients will be computed.

DESCRIPTION Uses the results of a transient analysis to compute the Fourier coefficients for the first nine harmonics. Note the `.FOUR` statement requires a `.TRAN` statement to perform the transient analysis.

EXAMPLE `.FOUR 10E3 V($N_0003)`

OUTPUT STATEMENTS

OPERATING POINT

SYNTAX `.OP`

DESCRIPTION Outputs information describing the bias point computations for the circuit being simulated.

EXAMPLE `.OP`

PRINT RESULTS

SYNTAX `.PRINT` type vname

where

type identifies the type of analysis performed to generate the data being printed and must be one of the keywords `DC`, `AC`, or `TRAN`;

vname is the variable name or the list of variable names for which values are to be printed.

DESCRIPTION Prints the results of circuit analysis in table form to an output file for each program variable specified.

EXAMPLE `.PRINT AC V(R_in) I(V_source)`

PROBE

SYNTAX `.PROBE` [vname]

where vname is the optionally specified variable or list of variables whose values from dc, ac, and transient analysis will be stored in the file that PROBE uses to generate plots. If no variable name is included, values of all circuit variables will be stored in the PROBE file.

DESCRIPTION Generates a file of data from dc, ac, and transient analysis used by PROBE to generate high-quality plots.

EXAMPLE `.PROBE V($N_0002) I(R_in) I(R_out)`

MISCELLANEOUS STATEMENTS

TITLE LINE

DESCRIPTION Each source file must have a title line as its first line.

EXAMPLE A Circuit to Simulate RLC Frequency Response

END STATEMENT

SYNTAX `.END`

DESCRIPTION Each source file must have an `.END` statement as its final line.

EXAMPLE `.END`

Index